Ethnobotany of

The Kondh, Poraja, Gadaba and Bonda

of the Koraput Region of Odisha, India

Ethnobotany of
The Kondh, Poraja, Gadaba and Bonda
of the Koraput Region of Odisha, India

F. Merlin Franco
D. Narasimhan

DK PRINTWORLD
Publishers of Indian Traditions

Cataloging in Publication Data — DK
[Courtesy: D.K. Agencies (P) Ltd. <docinfo@dkagencies.com>]

Franco, M. Merlin.

Ethnobotany of the Kondh, Poraja, Gadaba and Bonda of the Koraput region of Odisha, India / F. Merlin Franco, D. Narasimhan.

p. cm.

Includes bibliographical references (p.).

ISBN 13: 9788124606193

1. Kandh (Indic people) — Ethnobotany. 2. Parji (Indic people) — Ethnobotany. 3. Gadaba (Indic people) — Ethnobotany. 4. Bondo (Indic people) — Ethnobotany. 5. Ethnobotany — India — Koraput (District) I. Narasimhan, D., 1960- joint author. II. Title.

DDC 581.60954133 23

ISBN 13: 978-81-246-0619-3 ISBN 10: 81-246-0619-6
First published in India in 2012

The book has been published with the financial support from the Conservation Foundation, UK.

Published and printed by
D.K. PRINTWORLD (P) LTD.
Regd. Office : '*Vedaśrī*', F-395, Sudarshan Park
(Metro Station: Ramesh Nagar)
New Delhi – 110 015
Phones: (011) 2545 3975; 2546 6019; *Fax*: (011) 2546 5926
E-mail: indology@dkprintworld.com
Website: www.dkprintworld.com

Dedicated to
the traditional healers of the world

Preface

THIS book is based on a joint venture between Centre for Floristic Research, Madras Christian College and Integrated Rural Development of Weaker Sections in India (IRDWSI), a reputed non-profit organisation, working with the tribal communities of Odisha. The fieldwork was conducted between 2003 and 2009. Majority of the data were collected through mere observation, so as to present an insider's account of the tribal culture and its relationship with plants.

The first chapter is introductory in nature, providing an overview of ethnobotany, its history and importance in the modern world and the second chapter deals with the locality, communities and method of study. The methodology is explained in such a way that any reader could replicate it on the field. The third chapter lists the medicinal, cultural, culinary, economic and other uses of 210 species of plants and two sub-species, belonging to 77 families in alphabetical order. The local Odia, Kuvi, Poraja, Gadaba and Bonda names are also provided along with details of the knowledge holders who had provided the valuable information on them. The ethnotaxonomical knowledge of the four communities are provided in the fourth chapter, using Berlin et al. system of folk classification as the template. The chapter also discusses about the importance of local names, their origin, ecological significance, current status of the four tribal languages and implications for loss of languages.

The fifth chapter discusses about the plants used in the drug-based and spiritual healing processes. It elaborates the philosophies behind knowledge transmission such as (1) Divination, (2) Hereditary, (3) Discipleship and (4) Kinship. Disease diagnosis, diet restrictions and the rituals concerned are depicted in detail. The existence of an inverse relationship between the number of plant species used in drug-based healing and the number of species used in spiritual healing processes is explored for the first time. The sixth chapter on forests and NTFPs gives an insight about the effort put in by the communities in forest conservation, their land-use patterns, forest classification systems, list of NTFPs and their harvest-consumption patterns. The role of NGOs, middlemen and government agencies are also discussed.

The cultivars conserved by the communities, their agro-ecosystems and land management systems are presented in the seventh chapter. The cultural mechanisms such as festivals and their role in safeguarding agrobiodiversity of the region are depicted vividly. It calls upon the policy-makers concerned with addressing agrobiodiversity to take a multi-dimensional view than looking at agrobiodiversity as mere resources. The concluding chapter documents the different indigenous technologies involved in the processes of oil extraction, brewing and distillation.

Throughout the book, emphasis has been laid on the philosophical relationship of the communities with their ecosystem. The reader would therefore find that the book is essentially a good resource material on the biocultural diversity of the communities. Considering the holistic nature of the study undertaken, we are sure that this book would prove to be a valuable handbook for the tribal communities, policy-makers, academicians and social workers looking forward to accompanying the communities in their journey towards holistic, ethno-sensitive development.

F. Merlin Franco
D. Narasimhan

Contents

I keep thinking that the best ethnobotanist would be a member of a cultural minority and trained as a botanist and as an ethnologist, would study, from within and as a part of it, the traditional knowledge, cultural significance and the traditional management and use of the flora.

— Barrera 1972, cited in Toledo 1995

Acknowledgement

THIS book was possible only because of the infrastructural facilities and partial financial support extended by IRDWSI, Semiliguda, Odisha, for my research. I thank Dr William Stanley and Dr Sashi Prabha for their support and encouragement. I also thank my parents T. Francis and B.K. Pushpavally for partially sponsoring the research that forms the basis for this book. I also thank my siblings — the Merlins for their constant support.

I thank The Conservation Foundation, U.K., for their generous financial support without which this book would have remained a dream. Special thanks to David Shreeve, Director and Co-founder of the Foundation and Bill Moloney, Company Secretary of the Foundation for their special interest in this project and encouragement.

The University of Madras has warmly permitted us to base this book on my Doctoral dissertation. I thank the Vice-Chancellor, Registrar, Controller of Examinations, the Syndicate and others for their kind co-operation and encouragement.

Many thanks to Dr Alexander Mantramurti and Dr V.J. Philip, former principals, Madras Christian College, Dr R.W. Alexander Jesudasan, principal, Madras Christian College, Dr C. Livingstone and Late Joel Christopher, former heads and H.K.P. Devadoss, Head of the Department of Botany, Madras Christian College, for providing me with the facilities to carry out the research in the department.

The moral support and guidance provided by Dr P. Dayanandan, former Head of the Department of Botany, Madras Christian College and Thiru. Lawrence Surendra, sharpened my ideas. My sincere thanks to their noble hearts.

The catalytic support provided by Mohan Hontal and family, Gideon Soren and family, Madab Dalapati, Bilay Kumar Sen and family, Solomon Tunglut,

Dayadhan Suna and Marshall lessened the difficulties faced during the course of study and stay at Semiliguda.

I thank Mr and Mrs V. Sudhakar, Prabhat Kosla, Ms Pramila Kosla, Benudhar Sutar, Prakash Kora, Vinod Kosla, Manoranjan Patnaik, Sukhra Chellan and Mrs Hemalatha Hontal for their support.

I thank the knowledge holders: Guru Sillaiguda of Bonda Hills, Mrs Sonai Pangi of Chikalmari, Raghu Jani and Mrs Jamuna Pangi of Panasput, Prem Patro Jani and Ms Subha Dishari of Putisil, Hori Hontal of Kondh Pungar and Bondu Doli of Khojuriguda for sharing their valuable knowledge with me and also for the warm, unforgettable hospitality extended to me throughout the study. Their love made me feel at home throughout the study for which I shall ever remain indebted.

I sincerely thank Mrs Anne Dayanandan for her encouragement and critical comments on my manuscript.

I am thankful to Thiru. A.V. Balasubramanian, Dr K. Vijayalakshmi and staff of Centre for Indian Knowledge Systems, Kotturpuram, Chennai, for the help received towards the final phase of the study.

I thank Dr M.D. Muthukumaraswamy, Director, National Folklore Support Centre and the Librarian for permitting me to use their library. I also thank Dr A.G. Pandurangan, TBGRI, Thiruvananthapuram, Thiru. Santhosh Patnaik, ODAF, Bhubaneswar, Dr S. Amerjothi, former Head of the Department, Presidency College, Dr Rukmani Nirmal, Chettinad Hospital and Ms Padmini Nirmal for every help and encouragement received.

My sincere thanks to Dr G.A.I. Ebenezer, Dr M. Balusami, Leslie Lawrence, Dr W. Aris Dason of Department of Botany, Madras Christian College, Dr Nirmal Selvamony, Department of English, Madras Christian College, Samuel Sukumar, Warden Bishop Heber Hall, Madras Christian College, K. Vishwanatha, Department of Languages, Madras Christian College and Dr P. Ravichandran, Sri Paramakalyani Centre for Environmental Sciences, Manonmaniam Sundaranar University for their constant encouragement and meaningful discussions that helped me better my dissertation.

I thank Dr Raghupathi Subramanyam, University of Guelph, Canada, for sharing a few valuable research publications.

Special thanks to Dr Sheeba J. Irwin for always being there as a source of support and encouragement.

I am fortunate to have good friends in Dr Daliya George, Dr Sarah John, Luke Simon, J. John Cyril, John Jebaseelan, Mrs T.D. Nirmala Devi, Dr Rayson K. Alex, Mrs Susan Deborah, K. Krishnadhas and G. Manikandan which ensured the much needed moral and physical support for completing this book.

All thanks to my current and former colleagues at the Centre for Floristic Research: Dr Sherie Jesulyn David, Dr R. Sumathi, Mrs C. Chandrakala, Dr J. Jayanthi, Dr K. Karthikeyan, J. Soundarapandi, P. Saravanan, Mrs A.K. Rathnakumari, S. Thambidurai, Joseph Daniel, Ms R. Sheela Beryl Rajakumari, Ms K.M. Koperundevi, Mrs Josphin Mini, G. Gnanasekharan and N. Dhatchanamoorthy for their support and encouragement.

I recall with thanks, the timely help rendered by Mrs Chitra Porselvi, Mrs Valarmathi, Sureendar Raman, Raghunath Velankar and Mrs Anshu Singh for the successful completion of the study.

I owe my life to the good souls who cared to stop by and rescue me from an accident at Pottangi, Koraput. They preferred to stay anonymous. Yet, their kind gesture shall always be remembered with thanks.

I thank Dr D. Narasimhan for supervising the Doctoral study that forms the basis for this book and also for agreeing to co-author this book with me. He has been a good friend and a constant motivator for me. I also thank his family members for their constant support and encouragement.

F. Merlin Franco

1

Introduction

> What sets worlds in motion is the interplay of differences, their attractions and repulsions. The idea of a single civilisation for everyone implicit in the cult of progress and technique, impoverishes and mutilates us. Every view of the world that becomes extinct, every culture that disappears, diminishes a possibility of life!
>
> — Stephens et al. 2006

EVOLUTION of humankind is closely connected to the environment. As Ramakrishnan (2005) rightly points out, epistemology of early human societies as well as many of the present-day tribal and rural communities developed due to their close interaction with their ecosystems. Such ecological knowledge of these societies has facilitated their harmonious living with nature that is expressed in various forms such as beliefs, rituals, cultural practices and uses of biodiversity. The urban societies facilitated by modern technological advancements, have deviated from this, resulting in large-scale exploitation of nature. Hence, understanding the ecological knowledge of tribal and rural societies is necessary to conserve and sustain natural resources. The study of ecological knowledge of any society or community of people is called as ethnoecology and the branch of ethnoecology that deals with knowledge of people on plants is known as ethnobotany (Martin 1995; Davidson-Hunt 2000). Though ethnobotanical studies can involve any group of people, it is generally focused on tribal and other indigenous communities owing to the abundance of ethnobotanical knowledge among them.

Concept of Tribe

Tribal communities have mostly remained isolated from the mainstream communities, retaining their traditional ways of living. But the mainstream communities have seldom been sympathetic to the traditional culture and lifestyle of tribal communities. This apathy towards tribal culture has given a negative connotation to the very meaning of the word "tribal". The *Oxford English Dictionary* (2002) defines a tribe as a "social division in a traditional society consisting of linked families or communities with a common culture or dialect". However, it also recognises the derogatory meaning attached, based on the past attitude of the colonialists towards the so-called "primitive" or "uncivilised" people. Hence it advocates the usage of alternative terms such as "community" or "people". The International Labour Convention uses the word "Indigenous Communities" to describe tribal and other native communities, which is also a highly debated one (Guenther 2006; Barnard 2006). International Labour Organisation defines Indigenous People as

> peoples in independent countries who are regarded as indigenous on account of their descent from the populations which inhabited the country, or a geographical region to which the country belongs, at the time of conquest or colonisation or the establishment of present state boundaries and who, irrespective of their legal status, retain some or all of their own social, economic, cultural and political institutions. — Indigenous and Tribal Peoples Convention 1989

The Indian Constitution uses the term "Scheduled Tribes" to denote tribal communities. The term is believed to have a fixed meaning and denotes the enumeration of tribes. But the word "tribe" no more denotes political unit as of olden days, but refers to a group of people identified with poverty and backwardness. Hence the tribes of India prefer to be called as *ādivāsīs/ādibāsīs*, which, they believe, would help in maintaining their separate identity (Verma 1990). The term *ādivāsī* means "first settlers" and scores better than all other terms used in India such as *vanya jāti* (castes of forest), *vanavāsī* (inhabitants of forest), *pahāṛī* (hill dwellers), *ādimjāti* (original communities), *janajāti* (folk people) and *anusūcita janajāti* (scheduled tribe) (Vidyarthi and Rai 1976). The term *ādivāsī* is nearly identical to the term "First Nations" used by Canada to denote indigenous communities (www.afn.ca). However, throughout this study, we have used the term "tribal communities" in lieu of *ādivāsīs* as this is the only phrase used in official and legal circles.

Political Understanding of Tribal Communities

The economic and development indices of all tribal communities are not the same. Hence the Government of India has classified several tribal communities as *primitive,* aimed at providing special focus on development of these communities. The criteria used for notifying a community as primitive are

(i) Pre-agricultural level of technology,

(ii) Extremely low level of literacy,

(iii) Very low birth rate/stagnant population, and

(iv) General backwardness due to seclusion, and consequential archaic mode of living.

Based on the above criteria, 72 Scheduled Tribe communities were identified as *primitive* in 1960 by the Ministry of Home Affairs. In Odisha, there are 13 "primitive" tribal communities, viz. Birhor Bondo, Chuktia Bhunjia, Didayi, Dongria Kandha, Hill Kharia, Juang, Kutia Kandha, Lanjia Saora, Lodha, Mankirdia, Paudi Bhuyan, and Saora (Behura & Mohanty 1998; Devi 2002). However, the term "primitive" is derogatory and there is a necessity to look for alternate terms when describing a community which has taken pain to stick to their traditional ways of living. Mahapatra (1972) is of the view that any society can produce superior values and to brand a community as "primitive" by comparing its technological possessions with those of the mainstream communities is not healthy. He says that the communities that are "primitive" on the basis of the technology tend to have superior values, codes of conduct and social ideals. Behura and Mohanty (1998) advocate the use of the term "Vulnerable Ethnocultural Groups" (VEGs). They justify the use of the term "vulnerable" because of the several restrictions placed on their access to their own natural resources thereby depriving them of autonomy and food security.

History and Evolution of Ethnobotany

Ethnobotanical knowledge has evolved long before writing came into being. Archaeoethnobotanical data available as of today reveals the extensive knowledge possessed by ancient societies. Summer and winter crops at all processing stages in Harappan Civilisation were excavated and studied by David and Kramer (2001) to give an understanding of the Harappan food culture. Sculptures of plants available in temples and monuments give an idea of plant resources then used. Ixion's flower depicted in Ixion's wheel as per Greek mythology is now identified as *Datura stramonium* (Ruck and Staples 2001). Gupta (1996) provides an exhaustive list of plants depicted in Indian temple art. The depicted plants include species such as *Mangifera indica, Anthocephalus cadamba, Ficus benghalensis* and *Aegle marmelos*. The knowledge on plants that was traditionally preserved and transmitted in the form of paintings, sculptures, oral songs and poems was later written down in the form of literature (Sensarma 1998). This gave birth to the Indian codified systems of traditional knowledge. The traditional ethnobotanical knowledge documented in the Indian codified systems such as Āyurveda, Siddha and Unani is believed to have come from the contemporary ethnobotanical knowledge, compiled by various authors over a period of time. Ethnobotanical work in China was first recorded in the Chinese herbal *Pen ts'ao kang mu* written by Li Shih-chen (Schultes and von Reis 1995). In Greece, ethnobotanical work was pioneered by Theophrastus and Aristotle. Aristotle's teachings had influenced much of the European herbals and hence Aristotle is considered as the father of Western ethnobotany. However, even before the times of Aristotle and Theophrastus, the Greeks were good ethnobotanists. This could be understood from the association of plants such as *Papaver somnifer, Papaver rhoeas* and *Amanita muscaria* in Greek mythology. In CE 77, Dioscorides, the Greek surgeon in the court of Nero compiled a Materia Medica which listed the uses of about 700 plants (Davis 1995).

The term "ethnobotany" was coined by the American Ethnobotanist — John William Harshberger — in 1896. He stated that "Ethnobotany was the first science that originated with the evolution of existence of human beings on this planet." Ethnobotany is defined in various ways. Schultes (1962) defines it as the "use of plants in primitive societies". Schultes and von Reis in 1995 modified this definition so as to mean "the study of human evaluation and manipulation of plant materials, substances, and phenomena, including relevant concepts, in primitive or unlettered societies". According to Jain (1986) "ethnobotany implies a total interrelationship between human beings and plants." Jain again in 2001 modified his definition for ethnobotany as "the study of total natural and traditional interrelationships between man and plants, and man's domesticated animals". Of all these interpretations, that of Jain (1986) is the most appropriate as it presents the complete meaning of ethnobotany.

Role of Ethnobotany

Very often, ethnobotany is equated with medicinal botany. With the advent of bio-prospecting, this has further advanced where ethnobotanists are seen as mere couriers who ferry knowledge from the traditional domain to the scientific domain.

This also highlights the attention that traditional knowledge is currently receiving. Traditional knowledge and medicines immensely contributed to the well-being of humans till the end of Second World War, that saw the rise and rapid growth of allopathic system of medicine resulting in marginalisation of traditional medicines (Naranjo 1995). Yet, local communities in Asian, African and Latin American countries depend on traditional medicines for their health care requirements. Gadgil and Rao (1998) estimate that around 70 per cent of the population in India is dependent on traditional plant-based medicine. This is possible only due to the presence of an estimated 700,000 traditional healers and midwives (Sadgopal 2008). Around 60 per cent of the population in Hong Kong consults traditional doctors whereas 35 per cent of the outpatients and 22 per cent of the inpatients in rural China are treated with traditional medicines (Natesh 2001). Of the world's 25 top-selling pharmaceuticals, 10 were originally sourced from animals, plants or micro-organisms. In 1995, these accounted for nearly US $ 14 billion in global sales (McGhee 2003). Apart from the large quantity of traditional plant-based medicine used, there are a number of drugs derived from plants. Bioprospecting uses the "clues" rendered by traditional knowledge to derive new therapeutic compounds. Kumar and Tarui (2003) have proposed a mathematical model based on earlier models to analyse how indigenous and traditional knowledge influences the process of bioprospecting. Based on their model, they have concluded that if the bioprospecting search is based on ethnobotanical information available from the local community, then the value of bioprospecting benefits will be higher.

Ethnobotany and Understanding of Indigenous Science

Traditional knowledge is often undermined and generalised as outcome of trial and error experiments. This underestimates the local knowledge on natural processes and can be viewed as an extension of the human ethnocentrism. Traditional knowledge has to be understood from the local communities' point of scientific understanding. There are a number of traditional scientific processes recorded which couldn't have been achieved but for the immense knowledge and understanding of the local community. The simple process of liquor-brewing is one example of the traditional understanding of organic chemistry where liquor can be brewed virtually out of any carbohydrate containing food. Davis (1995) explains the various complex steps involved in the production of the dart poison curare from *Chondrodendron spp.* and the entheogen *ayahuasca* from *Banisteriopsis caapi.* Davis has shown that the process leading to the above products could not have arisen out of trial and error but from proper understanding of chemical processes occurring in nature.

Ethnobotany and Culture

Ethnobotany aims at studying the entire dynamic relationship between plants and human beings, existing both at the community and personal levels. Religion and culture are two important media that facilitate the interaction of the community and individual with nature. Several plants that deeply associated with religious practices, often to induce visions have been studied ethnobotanically. Such plants are now known as "Entheogens" meaning "God generated within", a term coined by Ruck and his co-workers (Wasson 1986; Ruck and Staples 2001). Several entheogens have been used till today of which the most popular one from the Indian context is the *soma*, mentioned in *Ṛgveda*. Wasson has identified the source of *soma* as *Amanita muscaria*, a Basidiomycetes fungus. It has been used under the name *Kakulja Hurakan* meaning thunderbolt by the Maya who use it even today as an entheogen. Other fungi too were extensively used throughout the world for inducing divine visions. *Claviceps purpurea*, commonly known as ergot could have been used as an entheogen by the hierophants of Eleusis. The Mexicans use *Psilocybe caerulescens* and *P. mexicana* as entheogens. In India, *Stropharia cubensis* another Basidiomycetes fungus is believed to be used as an entheogen by the Santal and Ho of Similipal area (Wasson 1986).

Culture is the medium for traditional knowledge conservation and transmission. Festivals, rituals and customs are cultural elements that ensure the survival of traditional knowledge (Eyssartier et al. 2008; Napoli 2008). Much of the traditional knowledge available today could not have survived but for the important role played by culture. The key position enjoyed by plants in human culture is also reflected in folklores. Hence, study of folklore entities such as nomenclature, songs, proverbs, sayings and myths throw light on otherwise hidden knowledge on plants. There are numerous instances where folklores have revealed priceless knowledge of the communities (Austin 2007). Besides revealing the current plants-people relationship, folklores, owing to their antiquity, may also reveal information about the historical uses of plants, their cultivars and related extinct practices which is of great interest to the field of ethnobotany (Tsehaye & Kebebew 2006). Traditional conservation practices such as sacred landscapes, conservation through cremation sites or graveyards, sacred trees, etc. are also cultural elements extensively studied in ethnobotany (Gadgil & Vartak 1975; Dafni et al. 2006). However, very few ethnobotanical studies give importance to a holistic understanding of the relationship between human culture and plants. Hence, one might be tempted to see the emerging field of "Biocultural diversity" as a panacea.

Ethnobotany and Traditional Conservational Practices

Ethnobotany can play an important role in promoting understanding of traditional methods of biodiversity conservation and sustainable utilisation of resources (Prance 1997). Societies all over the world have been judiciously using the available resources, facilitated by the traditional conservation practices such as sacred groves, sacred trees and sacred pools. Sacred groves are "traditionally managed" forest patches that functionally link social life and forest management system of a region (Boraiah et al. 2003). Maikhuri et al. (2003) describe this as a manifestation of nature-ephemeral perception of poor resilience capacity of natural ecosystems. According to them, such a perception arises out of fear for utilisation of natural resources or out of a belief that conservation of any ecosystem is vital for a sustainable livelihood. The exploding human population has been exerting a huge pressure on the ecosystems and also hampering the conservation efforts. Hence traditional systems of conservation such as sacred sites have gained much importance today.

Malhotra et al. (2001) has attempted to provide an understanding of the total number of sacred groves existing in India (Table 1). However, it does not represent a near complete survey and requires updating. For instance, King (2005) has studied 252 sacred groves in Kolli hills alone, a small hill range in Tamil Nadu. Hence, the total number of sacred groves in the State of Tamil Nadu may actually be far higher than indicated. Sacred groves conserve several plant species and animals, serve as water-sheds, seed banks, and also play a great role in maintaining the cultural values of the community (Raghavendra et al. 2005; Sashikumar 2005; Boraiah et al. 2003; Kunhikannan 2005; Marglin & Mishra 1993). Hence they are hailed as excellent models of conservation, governed by strict customary laws and taboos (Lingard et al. 2003). It is the belief that forms the cornerstone of such traditional methods of conservation. It is a common observation that those species which are protected by taboos are either keystone species or have greater cultural value.

Table 1: Reported Number of Sacred Groves in India

State	No. of Sacred Groves
Andhra Pradesh	750
Arunachal Pradesh	58
Assam	40
Chhattisgarh	600
Gujarat	29
Haryana	248
Himachal Pradesh	5000
Jharkhand	21
Karnataka	1424
Kerala	2000
Maharashtra	1600
Manipur	365
Meghalaya	79
Odisha	322
Rajasthan	9
Sikkim	56
Tamil Nadu	448
Uttaranchal	1
West Bengal	670
Total	**13720**

Source: Malhotra et al. (2001).

The aim of this book is to provide an overview of the ethnobotanical knowledge of the Kondh, Poraja, Gadaba and the Hill Bonda communities of the Koraput region of Odisha, India. The reader would be able to understand that ethnobotany has been applied in a holistic sense, so as to depict the prevailing biocultural diversity. The study area comprises the political districts of Koraput and Malkangiri and all the four communities studied have been notified as tribal communities by the Government of India.

2

Locality, Communities and Method of Study

Odisha

Odisha (Orissa) lies between 81° 24' and 87° 29' East Longitudes and 17° 48' and 22° 34' North Latitudes. It is bound by the State of Bihar on the north, Madhya Pradesh on the west, Andhra Pradesh on the south, West Bengal on the north-east and by the Bay of Bengal on the south-east. The total geographic area of the State is 155,707 sq km (Saxena & Brahmam 1994).

Odisha was a division of Bengal Presidency till 1901. In 1922, Odisha and Bihar were separated from Bengal Presidency to constitute separate provinces of Bihar and Odisha. On 1 April 1936, Odisha was accorded the status of a separate province with six districts. During the formation of this separate province, certain portion of the agency tracts of Visakhapatnam of Madras Presidency were also annexed so as to form the Koraput district. On 1 August 1949, 24 Princely States were also merged with Odisha.

On 1 April 1993, 14 new districts were carved out increasing the total number of districts to 27. Three more districts were created subsequently increasing the total number of districts to 30 (Saxena & Brahmam 1994). Until August 2009, the State of Odisha was known as Orissa. An assembly resolution reverted its name to its original form — Odisha — along with the name of language reverted to Odia, from Oriya (*The Times of India*, 2009).

According to the Census of India 2001 the total population of Odisha is 36,804,660. Of this, the tribal population is 8,145,081, which comes to about 22.1 per cent of the total population making it one of the tribal-rich states in India. Of the total 30 districts, 12 are designated as Tribal districts. The names of the tribal districts and their population data are given in Table 2. An outline of the state of Odisha is shown in Map 1.

Table 2: Tribal Districts and Their Population

District	Geographic area (km²)	Total cover (km²)	Total population	Tribal population	Percentage of Tribal population
Baleshwar	3,806	309	2,024,508	228,454	11.3
Gajpati	4,325	2,484	518,837	263,476	50.8
Kalahandi	7,920	2,260	1,335,494	382,573	28.6
Keonjhar	8,303	3,236	1,561,990	695,141	44.5
Khandamal	8,021	5,451	648,201	336,807	52.0
Koraput	8,807	1,557	1,180,637	585,830	49.6
Malkangiri	5,791	2,212	504,198	289,538	57.4
Mayurbhanj	10,418	4,005	2,223,456	1,258,459	56.6
Nawarangpur	5,291	1,123	1,025,766	564,480	55.0
Rayagada	7,073	3,061	831,109	463,418	55.8
Sambalpur	6,657	3,288	935,613	322,770	34.5
Sundargarh	9,712	4,025	1,830,673	918,903	50.2

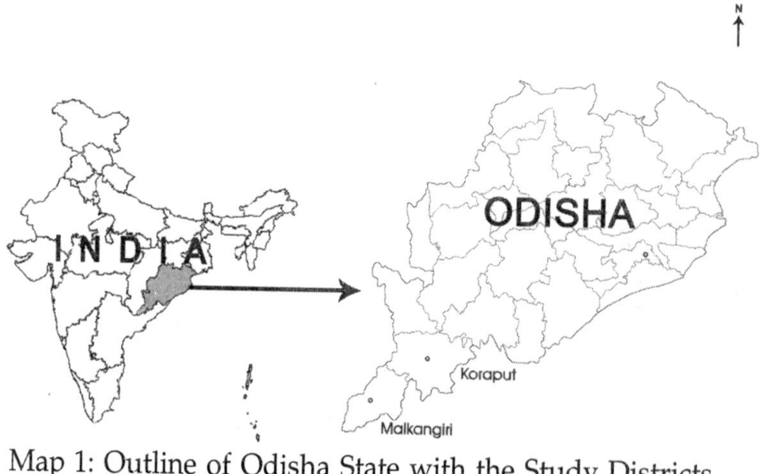

Map 1: Outline of Odisha State with the Study Districts

There are 62 tribal communities in the state of Odisha (Mandal et al. 2002). Of these, 14 speak Dravidian languages, 22 speak Muṇḍa and 26 speak Oḍiā. The different tribal communities and the language spoken are provided in Table 3.

Table 3: Tribal Communities of Odisha and Their Linguistic Affinities

Munda	Odia	Dravidian
Bhumija	Bagata	Chenchu
Birhor	Baiga	Dal
Bondo Poraja	Banjara/Banjari	Dharua
Desua Bhumija	Bathudi	Gandia
Didayi	Bhottada	Gond/Gondo
Gadaba	Bhuiya/Bhuyan	Jatapu
Ho	Bhumia	Kandha Gauda
Juang	Bhunjia	Kondh/Kandha
Kharia	Binjhia/Binjhoa	Kisan
Kol	Ghara	Konda Dora
Kolah Lohara	Holva	Koya
Kolha	Kawar	Madia
Kora	Kharwar	Oraon
Korua	Koli/Malhar	Poraja
Lodha	Kotia	
Mahali	Kulis	
Mankidi	Matya	
Mankirdia	Omanatya	
Mirdha	Pentia	
Munda/Munda Lohara	Rajuar	
Mundari	Sounti	
Parenga	Tharva	
Santal		
Saora		
Shabar Lodha		

Note: Gadaba, who are considered as of Austro-Asiatic stock, speak Gutob of Muṇḍa group. But the Ollar Gadaba of Koraput district speak Ollari, which is Dravidian and hence not included in the above Table.

Source: Nayak (1990).

Koraput District

Koraput district, formed on 1 April 1936 derives its name from the city of Koraput which is its headquarters (*figs* 1-4). There are several theories regarding the origin of the name Koraput. Bell (1945) is of the view that Koraput has been derived from Kora-puṭṭi meaning "the hamlet of nux-vomica". The second theory says that Koraput originated from Karakapentho (hailstone). The third theory says that a man named Khora Naik laid the foundation for the village. And the village came to be known as Korapūṭu which later got abbreviated to Koraput (Senapati & Sahu 1966). When formed, the total area of the district was 10,498.95 sq miles. It was the largest district in Odisha then. With effect from 1 April 1993, the district was divided into four districts, viz. Koraput, Malkangiri, Nawarangpur and Rayagada.

The present-day Koraput district falls between 17° 40' – 20° 7' North Latitude and 81° 24' – 84° 2' East Longitude. According to Forest Survey of India (FSI) (2003), the total geographic area of the district is 8,807 sq km. The district borders are formed by Rayagada and Srikakulam (A.P) districts in the east, Bastar district of Chhattisgarh in the west, Nawarangpur district in the north and Vizianagaram and Visakhapatnam districts in the south. The highest peak of the state called *Deomali parbat* (1673 m) also called as *Deomadi* by the tribal communities is in the Koraput district.

FORESTS AND VEGETATION

In 1967, about 70 per cent of the district was under forest cover (Senapati & Sahu 1966). But in 1990, it had diminished to less than 30 per cent (Malhotra et al. 1998) and by 2003, it had further declined to 17.68 per cent, i.e. 1,557 sq km, of which moderately dense forests form 729 sq km and open forests form about 828 sq km (FSI, 2003). According to Champion and Seth (1962), the forests of Koraput district fall into three types, viz. tropical semi-evergreen forests, *Terminalia tomentosa* forests of north Indian moist deciduous forests type, and dry deciduous scrub of tropical dry deciduous forests type. However, secondary forests growing on lands which were degraded earlier due to anthropogenic activities such as shifting cultivation and deforestation, show varied characters and are dominated by scrub elements.

Malkangiri District

Malkangiri existed as a subdivision of the erstwhile undivided Koraput district from 1962 to 1992. On 2 October 1992, as per the Government of Odisha Notification No. 49137/R, dated 01-10-1992, the present Malkangiri district was formed. This came into effect from April, 1993 (Saxena & Brahmam, 1994).

According to FSI (2003), the total geographic area of Malkangiri district is 5,791 sq km. It lies between 17° 45' N to 18° 40' N Latitude and 81° 10' E to 82° 00' E longitude. Malkangiri is bound by the Koraput district and the Visakhapatnam – East Godavari districts of Andhra Pradesh in the east, Bastar district of Chhattisgarh in the west, Koraput district in the north and East Godavari – Khammam districts of Andhra Pradesh in the south. The forests found are mainly of northern semi-evergreen forests type as per the Champion and Seth classification of 1962. Owing to its recent origin, not much literature is available on Malkangiri district.

Selection of Communities and Sampling

The tribal communities of Koraput region were always at the receiving end when it comes to their relationship with the mainstream people. The various bad experiences of exploitation had inflicted wounds in their minds which made them view every outsider sceptically. Fortunately, IRDWSI, a non-governmental organisation (NGO) working with the people of the region had a good reputation among the tribal people for their selfless service. Considering the fact that it has never been easy for an outsider to gain the confidence of the communities in this region, we decided to get introduced to them through IRDWSI. With the help of IRDWSI, we conducted a reconnaissance survey in 25 villages of the Koraput region during November–December 2003. Though we had initially planned to work with the Kondh alone, the reconnaissance survey changed our idea and helped us to understand the diversity of traditions and knowledge. This made us revise the original plan and a proposal for studying the ethnobotanical knowledge of four tribal communities, viz. Kondh, Poraja, Gadaba and the Hill Bonda, two each from the Dravidian and the Austro-Asiatic linguistic groups was developed (*figs*. 5-16). A brief description of the communities selected for the study is provided below.

Kondh

Kondh belong to two linguistic groups, Kui and Kuvi, both belonging to the Dravidian family (Banerjee 1969). The speakers of Kui are called as Kuttia Kandha and inhabit the Kandhmal region. Kuvi is spoken in the undivided Koraput district by two other sections of Kondhs, namely the Dongria Kandha of the Niyamgiri hills in the Bissam Cuttack region and the Plains-Kandhas who inhabit the plains of the Rayagada district. Banerjee also records the existence of two other groups who go by the name Jatapu and Kondadora, who are considered by him as the offshoots of the Kondh. However, Senapati and Kuanr (1984) list the Kondadora as a separate tribe. The Dongria Kondh consider the Niyamgiri hills holy and were recently in the news for vehemently fighting against a mining company that sought to displace them from their habitat, besides desecrating the Niyamgiri hills (Das 2010). Nayak et al. (1990) consider the Phulbani district as the centre of Kondh territory. According to them, there are five sub-groups of Kondh: (i) Maliah Kondh of Phulbani district who speak Kui, (ii) Kuttia Kondh of the south-west Phulbani district who speak Kui which differs from the Maliah Kondh, (iii) Dongria Kondh of the Niyamgiri hills, (iv) The Kuvi of the north Koraput district who speak Kuvi which is extremely nasal, and (v) Desia Kondh who are migrant groups of Maliah Kondh and are found below the Eastern Ghats, coastal plains, central Kalahandi, Bolangir and Dhenkanal district. The people of the Kondh community of the present study prefer to be called as Kuvi Kondh.

The origin of name Kondh is a subject of debate. Dalton (1872), Sahoo (1996) and Misra (2002) prefer to call them as Kandhs. Aiyappan (1948) calls them as Khond. Chaudhury (1989), Jena et al. (2002) and Jena et al. (2006) refer to them as Kondh. Throughout this study, we refer to the community as Kondh. Their population in 1961 was about 79,034 in the undivided Koraput district, excluding the Kashipur tahsil. Males numbered 39,875 and females 39,159 (Senapati & Sahu 1966). In 1971, the population increased to 325,144 (Senapati & Kuanr 1984). In 1991, the total population in the undivided Koraput district was 414,206 (Satpathy 1992).

ORIGIN OF KONDH

The Kondh believe that they are the descendants of a couple who inhabited the earth when the earth was filled with water. As there was no land available, they were living in a *ḍoṅgā* (boat) carved out from the *Lekho* (*Bombax ceiba*) tree, locally known as *Simili gosch*. Hence the boat was called as *Lekho ḍoṅgā*. It was the children of this couple who later branched out into the whole tribe (Jena et al. 2002). Even today, the *Lekho* tree is considered sacred and highly revered by the Kondh of the Koraput district. They also point out to the *Deomadi* peak (Odia: *Deomali*) and its nearby smaller cliff as brother and sister of the couple respectively. As a result, the *Deomadi* peak is considered sacred. By suggesting that the earth was filled with water, the tale shrugs off the possibility of existence of any other society, thereby establishing the antiquity of the community. Likewise, *Deomadi* being the tallest peak in Odisha is used to prove their might over other communities.

CULTURE

The Kondh live in hamlets at the foothills of mountains. These hamlets are mostly situated in inaccessible areas. At the entrance to each village, there is a meeting place — a raised platform made of stones called as *sadhur.* Unlike the *sadhur* of other communities in the region, the Kondh *sadhur* is not built near or around any trees. The houses are constructed mainly of mud, reinforced with bamboo and thatch. Nowadays, these patterns have changed with tiled-roof as well as cement-asbestos roofed houses replacing the huts. Though these houses are more durable, they do not provide insulation from heat and cold as the grass-tiled huts do. The houses consist of two rooms — a living room and a kitchen. Sometimes a *piṇḍa* (veranda) is also seen at the front. The construction of *piṇḍa* could be a recent one, introduced due to the various housing schemes done by various NGOs operating in the area. The houses are built in linear manner, in an east-west alignment (*fig.* 5). This pattern of alignment in the east-west direction is a characteristic feature of Dravidian societies; the earliest known being the Indus Valley Civilisation where the streets were built either in north-south or east-west direction to maximise wind flow (Lahiri 1970). For the Kondh, this practice maximises the reception of the sun's radiation equally by all houses which is essential to escape the clutches of cold during the months of November to February. In the Koraput region, the pattern of linear construction of houses is unique to the Kondh and provides them the liberty of having two entrances to their houses — a front one and a rear one. As a result of the existence of a rear door, most of the houses have a kitchen garden, at least rudimentary in nature. It is said that every Kondh village has separate dormitories for boys and girls (Patnaik et al. 2002). This system is not practised by the Kondh of the study area. This is a unique feature of the Kondh in the *Deomadi* area.

Traditionally, Kondh were considered as specialists in wooden comb-making. It is said that the Kondh men always used to sport a comb in their long, knotted hair. However, this practice has disappeared because of the popularity of modern hairstyles and the labour intensive nature of comb-making. Both men and women pierce their ears and noses and wear ornaments, mostly gold (Devi 1972). Even the middle-aged Kondh men of today who are not seen wearing nose rings have marks of nose piercing that was done during childhood. Kondh women sport tattoos on their body, just like any other tribe of the Koraput region. Solid round dots tattooed in a linear fashion in the forearms is unique to the Kondh women. They believe that when they decease and go to heaven, the almighty would hand them a large stinging *piḍika* (worm). However, on seeing the dots on the arms, the worm would get scared and wriggle off. The Kondh are also known for their elaborate material art. Their house doors, walls and Meriah poles showcase their artistic skills. Their house doors were once carved with images of humans and animals in sexual unison. This practice too is not seen these days. Dash and Pradhan (2006) provide an elaborate compilation on the arts and crafts of Kondh community.

The Kondh are an endogamous tribe with the clans strictly exogamous. The lineage is patriarchal. Marriages do not happen within a clan since members of a clan are considered as siblings. After marriage, the offspring adopt the clan of their father. As a result, the maternal uncle's daughter is considered as a potential spouse. Marriages are broadly of four types:

(i) Negotiated (*Mangini*): The boy or the girl finds/falls in love with a suitable partner and reports to their parents. The boy's parents then negotiate the marriage with the girl's parents. Liquor is offered to the girl's parents and its acceptance by her parents signifies the sanctioning of marriage. The bride price is then negotiated by the elders. There is a long interval between the negotiation and marriage. It is the duty of the boy's parents to take liquor and other items such as clothes to the girl's family very often. This window period also helps to bring both the families together and promote good understanding.

(ii) Marriage after eloping (*Utilia*): When a pair falls in love with each other, they elope with the help of friends. The act of elopement happens mostly on market days. It is the duty of the elders to accept and bring them back.

(iii) Out-of-wedlock love marriage (*Sagartha*): If one of the partners develops intimacy with a third person, then the person walks away from the wedlock. In this case, the new husband will have to compensate the former husband with the exact amount paid by him to his father-in-law while acquiring the bride.

(iv) Widow Remarriage: Widow/Widower remarriage is very commonly observed in the tribal communities of Koraput region. Rarely do we find a widow or a widower remaining single. If the husband happens to die, it is a normal practice for the wife to marry his younger brother.

Rarely, a fifth type of marriage where the boy and his friends stage a mock kidnap to take away the girl from the market may also take place. The parents then bring them home and the negotiations begin.

The *haat* or the weekly market plays an important role in the culture of all tribal communities in the Koraput region. Every village is attached to the nearby weekly market. The market is not only the place where goods — mainly the week's produce — are bartered or sold, but also the place where hearts meet. The *haat* is often mentioned in their folk songs and it is quite usual for a prospective lover to invite his sweetheart to the *haat* where he tries to woo her by getting her cosmetics and other gifts (*figs.* 15 & 16). Hence, when a young lad asks a girl if she is willing to accompany him to the *haat*, it is implied that he is asking for a date with the girl.

SOCIAL ORGANISATION

The head of the village is called as the *jani*, whose post is largely traditional. In the absence of an heir or when the *jani* is found inefficient, another one is elected, who then attaches the suffix *jani* to his name. It is said that the Kondh village head was originally known as *santa* and the name *jani* was adopted only later. Apart from the *jani*, there are three other ranks of people in the Kondh society — the *dishari, bejeyni* and *gurumani*. The *dishari* is the official astrologer of the village who also doubles as the herbalist, very often. He is also the caretaker of the village spiritual needs. He decides, conducts and performs the various activities concerned with the Kondh religion. The *bejeyni* are women who assist the *dishari* in his rituals (Patnaik 1972; Franco & Narasimhan 2006). Normally, there are four *bejeyni* in a village but the number may sometimes be as low as one to nil.

The system of *bejeyni* is unique to the Kondh society. It denotes the importance given to women in performing spiritual duties. The *bejeyni* assist or even carry out the rituals in the sacred groves. It is noteworthy that most of the tribal communities do no permit women to perform such duties and enter sacred groves. The system of *bejeyni* is not hereditary by default. But it is common to see daughters of *bejeyni* taking up their mother's role, only because of their acquaintance. Any woman willing to become a *bejeyni* has to serve as an apprentice to an elder *bejeyni*, learning the rituals and the chants associated with them. This is not an easy task considering the elaborate nature of rituals and the length of chants. Chants may continue for hours together. Hence the career of a *bejeyni* requires great deal of devotion and commitment. In the absence of a *dishari*, the *bejeyni* may lead the rituals.

The *gurumani* is the traditional birth attendant of the village. She takes care of delivery, post-natal care and the rituals associated with it. She is also involved in tutoring the young girls, teaching them the traditions, tales and customs associated with the community way of living.

KONDH RELIGION

The Kondh religion reflects their fear and reverence for nature and natural forces. Every action, both positive and negative as perceived by the human mind is believed to be the action of spirits, expressing themselves as various facets of nature. They are said to worship a number of gods and goddesses. Swain (1968) reports the worship of 84 deities on different occasions by the Kondh of Kanjamendi in Phulbani district. However, this is not observed in the Kondh of Koraput district. Though the Kondh here acknowledge the presence of a pantheon of gods, they are not aware of their identities and worship Tada Pennu (*Tada* = earth, *Pennu* = God) alone as God. Ancestral spirits are also worshipped at home. Unlike the neighbouring Hindu communities, Kondh offer cows as sacrifice and also consume beef.

Tada Pennu is worshipped for overall prosperity and for good harvest. Rituals are held before sowing and after harvest while animals are sacrificed to please the earth god. It is highly essential for the Tada Pennu to be satisfied with the highest grade of blood available on earth which gave rise to the Meriah sacrifice (Devasahayam 1970-71). Dalton (1872) describes in detail the Meriah sacrifices done to please a god Maniksoro ąnd the Earth goddess. The sacrifice of human beings and the subsequent offering of his blood to the earth is believed to fertilise earth and make it ready for agriculture. This sacrifice suggests that the Kondh are primarily agriculturists by profession. It should be noted that the name Maniksoro, mentioned in Dalton's records was not found to be in use or remembered in the study area. Swain (1968) is of the view that human sacrifice was carried out for three main reasons such as increasing the fertility of the soil, to strengthen the foundation of a building and to ensure good water supply from a well or pond. Elwin (1988) says that the desire to perform Meriah sacrifice was prevalent among the Kondh even in 1944.

Poraja

Poraja are found in the district of Ganjam and Koraput and speak a Dravidian dialect called "Parji", commonly referred to as Poraja bhasha (Rousseleau & Behera 2002-03). A few of them have also settled in Assam and West Bengal. According to Senapati and Sahu (1966), the Poraja occurrence in this region can be traced back to second century CE. They are also of the view that the Poraja are divided into 12 sub-tribes, each named after the region they inhabit. Singh (1998) recognises five sub-groups among the Poraja of West Bengal, viz. (1) Konda Poraja, (2) Jodia Poraja, (3) Pengo Poraja, (4) Pareng Poraja, and (5) Godwa Poraja. Aiyappan (1948) recognises seven subdivisions among the tribe as (1) Barang Poraja, (2) Pengu Poraja, (3) Khondi Poraja, (4) Parengi Poraja, (5) Munda or Bonda or Nanga or Langla Poraja, (6) Tagara Poraja, and (7) Kur or Didayi Poraja. According to Mohanty (1990), Poraja are of four types, viz. (1) Bada Poraja, (2) Bada Jodia Poraja or Pengo Poraja, (3) Konda Poraja or Chelia Poraja, and (4) Bareng Jodia Poraja. The population of Poraja in 1941 was 14,458 (Aiyappan 1948). In 1981, it was recorded as 267,185 in Odisha (Singh 1998). In 1991, the total Poraja population in the then undivided Koraput district was 322,545 individuals (Satpathy 1992).

Poraja are known by various names such as Paroja, Parja, Porja and Poraja. Senapati and Sahu (1966) and Singh (1998), refer to them as Paroja. Aiyappan (1948) calls them as Porja. However, Poraja of the study region pronounce it as "Poraja" and female elders some time pronounce it as "Porja". Hence we retain the name "Poraja".

CULTURE

The Poraja hamlets are situated at the foothills as well as in the valleys (*fig*. 6). The houses are not built in a linear fashion and do not show unidirectional alignment.

The houses mainly consist of two doors and a veranda. The absence of linear pattern of housing in the village permits the Poraja to have only one entrance to the house, at the front. This eliminates the chance for kitchen gardens. There is a *sadhur*, made of stones at the entrance to the village. The *sadhur* is mostly built around the *jodi* tree (*Ficus arnottiana*) or rarely around *Ficus religiosa.*

It is said that the kings who once ruled the region used to take away the Poraja girls by force. The frustrated womenfolk then found out a way to escape this. They started tattooing all over the face. Even today, the elder women of the Poraja communities can be seen with tattoos on their forehead. The Poraja have a great artistic tradition. Tattoos, drawings and paintings form an important part of this. Their artistic expressions are mostly composed of dots, circles and linear forms. Houses were once adorned with paintings and drawings throughout the year. Remnants of this can be observed even today during festivals (*figs.* 95 & 96). The Poraja were also once known for their folk songs (Franco & Narasimhan 2008). However, the songs are quickly disappearing along with their language.

Like Kondh, Poraja are also of patriarchal lineage. The clans are exogamous and the tribe endogamous. Marriages between people of the same village are strictly prohibited, as they are considered as brothers and sisters. The daughter of maternal uncle is a potential spouse. Child marriages between children of ages above 13 years were observed (*fig.* 11). However, it is a lesser evil when compared to the child marriages happening in the mainstream communities. The bride and her groom of Poraja start co-habiting only around the age of 18. Till then, they live in the same house and start developing affection for each other. The Poraja have six types of marriage such as negotiated (*mangini*), marriage after eloping (*utiliya*), out-of-wedlock love marriage (*sagartha*), kidnap (*jigani*), widow remarriage and *poisamundi* (marriage by girl's initiative). In *poisamundi*, a girl who falls in love with a boy enters his home and his parents are compelled to accept her as daughter-in-law. The system of paying bride price to her parents is practised and the rest of the processes are the same as that of Kondh.

Love marriage is mostly encouraged. Existence of the *dhangadi basa* (girls' dormitory) and *dhangada basa* (boys' dormitory) facilitates this (Ulaka 1976). The *dhangadi basa* and the *dhangada basa* are usually a widow and widower's house in the village respectively. Bachelors of the village who have a moustache are permitted to stay in the *dhangada basa* and spinsters who have attained puberty are permitted to stay in the *dhangadi basa*. Men from other villages visit the *dhangadi basa* and women from other villages visit the *dhangada basa* in the evenings and during festivals. These dormitories serve as the place where boys and girls are initiated into adulthood. They are taught livelihood lessons and also the fundamentals of the community's culture. The dormitory is also the place where dance, songs and folk tales are taught and also created. In due course of time, there is a strong bond formed between the inmates of a dormitory which in turn boosts the unity among villagers. The dormitories also promote understanding between villages of the same community. Besides, these dormitories are also institutions that ensure social security by acting as support mechanisms for the bereaved widows and widowers.

The Poraja celebrate 13 festivals: (1) *Poos porv,* (2) *Nondi porv,* (3) *Kandul baja,* (4) *Choith porv,* (5) *Kotna kavoni,* (6) *Bion jatra,* (7) *Gangamma jatra,* (8) *Azadu jatra,* (9) *Bondapano jatra,* (10) *Nua kavoni,* (11) *Dosra porv,* (12) *Dealli porv* and (13) *Pod mara porv.* Of these, *Dosra porv* and *Dealli porv* are considered as recent additions.

SOCIAL ORGANISATION

Family is the fundamental unit of the Poraja social organisation. The village headman is called as the *nayak*. The astrologer is also the native herbalist and the one in charge of religion is called as *dishari*. While the post of *nayak* is traditional, the post of *dishari* is not so. Any person who is well versed with the traditions, customs and strong in faith can be the *dishari*. The village also has a *gurumani* who is the traditional birth attendant and also takes care of minor spiritual needs. Every village has a platform of stones, called *sindhibor/sadhur* where the elders, including married women, meet and take decisions. The *sadhur* is usually built around or near species of Ficus such as *jodi* (*Ficus arnottiana*).

PORAJA RELIGION

The Poraja pantheon consists mostly of female goddesses except the Nondi Depta who occur as a pair. The other goddesses are Gangamma, Mauli Maa and Bimondi. Mud from termite hills is considered sacred and hence used in making forms of gods and goddesses. Poraja, like the Kondh, do not attach any religious significance to the cow and consume beef. The *dishari* is in charge of rituals. Besides this, families of certain clan are given specific worship tasks (Table 4).

Table 4: Poraja Clans and Their Religious Duties

Surname	Task Assigned
Doli	Worshipping Bimondi Maa. On rotation basis, a family has to keep Bimondi Maa in their home for three years.
Jani and *Maaji*	Worshipping Gangamma. They get the opportunity to have the goddess for eight days in a year.
Paadi	Worshipping Nondi Depta. They are in charge of Nondi Porv.

Gadaba

The name Gadaba is considered to have originated from the river Godavari in Andhra Pradesh or from Goda which means a brook in Odia (Thurston 1975; Singh 1998). This is corroborated by the Gadaba people of the study area. The word Gadaba is also said to denote a person who carries loads on his shoulders (Senapati & Sahu 1966). According to Singh (1998), Gadaba are of seven groups: Bodo, Parenga, Kalayi, Kapu, Kathivi and Jurunu. But Bell (1945) says that there are only three sub-groups in Gadaba, viz. Bado, Sano and Ollaro. Bado Gadaba belong to the Mundari linguistic groups while the Sano and the Ollaro Gadaba speak Ollari that belong to the Dravidian linguistic groups (Thusu & Jha 1972). In the area of study, the Gadaba distinguish only two sub-groups — the Bado and the San — and the Ollaro Gadaba seem to be a synonym for San Gadaba (Sabat et al. 1999). The Gadaba community studied, claim to belong to San Gadaba. The Gadaba are also one of the communities exploited for money by the mainstream communities in the Koraput region (Mohanty 1976).

CULTURE

The Gadaba are found settled in the foothills or in the valleys (*fig*. 7). Although their individual houses are constructed in a haphazard manner, the whole village gives a linear appearance from a distance. Houses are provided with only one entrance, at the front. The *sadhur* (platform of stones) is seen near the entrance to the village. Trees such as *jodi* are found in or near the *sadhur*. One unique feature of the Gadaba *sadhur* is the presence of separate male and female *sadhur*. The male *sadhur* is meant for the adult males and the female one for adult females. This feature is also noted in a few Hill Bonda villages.

SOCIAL ORGANISATION

The Gadaba are an endogamous tribe with clans exogamous. Senapati and Sahu (1966) are of the view that the tribal organisation is not rigid and any member of other superior castes such as Poraja, Bhatra or Muria can become a Gadaba at the expense of Rs. 2 or 3. However, the Gadaba of the study area are found to follow strict rules and are very much conscious of their identity. Baliarsing and Nayak (1996) provide a list of nine totemic groups and clans found among the Gadaba of the Lamtaput area. Table 5 gives the totemic groups of Gadaba.

Table 5: Gadaba Totemic Groups

Name of the Group	Elements Associated
Killo/Khila/Killa	Tiger
Hantal/Ontal	Snake
Khara/ Kara	Sun
Macha/Mach	Fish
Gollarie	Monkey
Pangi	Kite/Vulture
Balu	Bear
Sukri	Cow
Girem	Cat

Source: Baliarsing & Nayak 1996.

Baliarsing and Nayak (1996) also give a list of the title groups among Gadaba (Table 6). This holds good for the study area too.

GADABA RELIGION

The Gadaba worship a domestic god called Kew, which is in the form of a weapon, mostly a knife. This practice is slowly changing into a form of weapon worship (*āyudha pūjā*) with erosion in the belief that God occurs in the form of knife. They also worship the Goddess Gangamma. Sometimes, they are found to worship Nondi Depta in association with the neighbouring Poraja. Spirits are worshipped and pacified. It is a common belief among Gadaba that there are spirits residing on the ways leading to the village. Therefore, there are special rituals done to prevent entry of spirits into the village. There is a symbolic gate called *bejorna* at a considerable

Table 6: Title Groups of Gadaba and their Functional Roles

Title Group	Function in the Society
Bado Naika	One among this group serves as the administrative head of the village.
Kirsani	Assists the Bodo Naika in administration and act as the head in his absence.
Sisa	Village priest
Pujari	Assists the Sisa and is responsible for the preparation of food during festivals
Muduli	In-charge of revenue matters
Chalan	Messenger

Source: Baliarsing & Nayak 1996.

distance from the village beyond which exists the patrol range of the spirits (Berger 2001). Gadaba of the study area were beef-eaters till the state government enforced a ban on cow-slaughter, owing to pressure from Hindu communities.

Bonda

Bonda are members of Austro-Asiatic stock and are concentrated in Khairput block of Malkangiri district. They speak a language called Remo which also means "men". Elwin (1988) as well as Senapati and Sahu (1966) refer to them as Bonda while Singh (1998) and Patnaik et al. (1984) refer to them as Bondo. However these people prefer to call themselves as Bonda. Senapati and Sahu (1966) cite Henderson describing Bonda as one of the three classes of Gadaba. According to him, the Gadaba are divided into Bonda Poraja, Bonda Gadaba and Parenga Gadaba. Patnaik et al. (1984) classify Bonda on the basis of their location into three groups, viz.

(i) Hill Bonda or the Bara-Jangar group, consisting of 12 villages,

(ii) Bonda villages of the plains (lower Bonda), and

(iii) Gadaba group of Hill Bonda.

The Bonda live at the Khairput block of Malkangiri district in Odisha. The lower Bonda live on the foothills while the Hill Bonda live on the top of the *ghat*s, commonly called as Bonda hills or Bonda Ghat. The Gadaba group live at Onmal and Ongael of the Bonda hills. Mohanty and Behura (2005) have classified the "Bonda ecological niche" into four eco-cultural zones. They are:

(i) *Mudulipada zone*: Altitude of 400 to 900 m with open mixed jungle and cultural homogeneity,

(ii) *Andrahal zone*: Altitude of 600 to 1000 m. Open mixed jungle with better forest coverage. The culture is influenced by the neighbouring Gadaba people,

(iii) *Pindajangar zone*: Altitude of 500 to 900 m. Dense forest cover and cultural homogeneity seen, and

(iv) *Patraput zone*: Altitude of 500 to 800 m. Good forest growth. Fishing practised.

As per the 1941 census, the population of Hill Bonda was only 2,565 (Senapati & Sahu 1966) and in the 1981 census, their total population was about 5895 (Singh 1998). This study is focused on the Hill Bonda alone and hence, any usage of the name "Bonda" strictly refers to the Hill Bonda alone.

CULTURE

The Hill Bonda build houses out of mud, reinforced with split bamboo (*figs*. 8 and 14). The houses generally consist of one big hall, partitioned into two. Sometimes, there is a small veranda before the house. Pigsty or goat pen is found attached to the houses. There is a *sadhur* at the centre of the village, built near *Ficus arnottiana* and sometimes even *Artocarpus heterophyllus*.

Just like any other tribe in Koraput district, the weekly market or *haat* plays an important role in the Bonda culture. Their markets are situated at the foothills thereby facilitating exchange of goods with the lower Bonda and other communities, including traders. Unlike other communities who get access to the *haat* only once a week, the Bonda have seven weekly markets happening in a week at different places around Bonda Ghat. A list of the seven weekly markets is given in Table 7.

Villages are mostly of a single clan. The tribe is endogamous and clans exogamous. As a result, marriage between people of the same village is prohibited. Marriages within the same title group are permitted provided they are from different villages. Girls' dormitory called *silanidingo* and boys' dormitory (*lingerisung*) are present and they are houses of widow and widower respectively. In the Bonda community, usually the groom is younger to the bride by at least seven years. Marriages are of four types, viz. love marriage, widow remarriage,

Table 7: Weekly Markets of Bonda and the Products Traded

Day of the Week	Name of the Market	Specialities of the market	
		Sold/Exchanged	Purchased
Sunday	Mundiguda	All items	Clothes
Monday	—	—	—
Tuesday	Govindapally	All items	Cattle
Wednesday	Khairput	All items	Cattle
Thursday	Kudumalguma	All items	Goat
	Onkadelly	Olsi	Bithi (Salt)
Friday	Mathili	All items	Bamboo baskets
Saturday	Chotaguda	All items	Cattle

out-of-wedlock marriage and arranged marriage (*woiboi*). In arranged marriage, the girl is "booked" when she is eight or nine years and when she grows up, she is brought home. Bride price (*jhola*) has to be paid to the bride's father.

Table 8: Days of the Week and the Corresponding Bonda Child Names

Day of the Week	Name of the child	
	Male	Female
Sunday	Aadi	Aadiwari
Monday	Somba	Sombhari
Tuesday	Mongla	Mongli
Wednesday	Budha	Budhey
Thursday	Lochim	Gurwari, Lochmo
Friday	Sukra	Sukri
Saturday	Soniya	Shonki

The Hill Bonda has a unique way of naming their children. Their names are characteristic of the days on which the child is born (Table 8). The same practice is also observed in the Konda Reddi community of Eastern Ghats.

SOCIAL ORGANISATION

The tribe is divided into two clans: *Ontal* and *Killo*. Besides this, there are title groups which are assigned with different tasks (Table 9).

Table 9: Bonda Title Groups and Their Roles in the Society

Surname	Task assigned
Sisha	Considered to be a royal clan. The village ritualist is from this clan
Chalan	To carry out dealings with the *sarkar* (government)
Kirsani	Village administration
Muduli	Village administration
Bathiri	Village administration
Bodnaik	Village administration
Maantra	Responsible for rituals happening at the *Paatkonda*.

Every village has a leader called *nayak* who is generally elected once in three years by the villagers. The 12 villages of Bara Jangar group together form a confederation which functions under a head (the *badnayak*) who is elected by the village heads of these twelve villages once in three years on the day of *Magha* festival. He is in-charge of inter-village affairs and disputes.

BONDA RELIGION

The Bonda believe in the existence of numerous gods. It takes a pious, learned man to establish relationship with them. This relationship being an individual affair is strictly one-to-one and the names of such gods are not revealed. However, there are a few common gods. The most important of them is the *patkanda mahaprabhu*, who occurs in the form of a knife, hidden at his sacred grove called *patkanda* at Mudulipoda. He is the nucleus of Bonda life and culture. Besides *patkanda mahaprabhu*, there are three other village gods such as Bursing, Hundi, and Mouli. Bonda offer cow as sacrifice and also consume beef.

Methodology

The initiation and course of study is summarised below.

A detailed project proposal was submitted to IRDWSI along with the request to help us in carrying out the study. IRDWSI agreed to depute one of its staff members as facilitator-cum-interpreter during field trips to the village. We were also permitted to use the existing infrastructure of the NGO. For each community, four villages were selected as "core" villages for detailed study. These villages were selected on the basis of their remoteness from the National Highway, lifestyle and rapport with IRDWSI. Besides the core villages, visits and brief study were undertaken in other neighbouring villages. Such villages include Aligao, Pakhajola, Mattiaguda and Mojikundali of the Koraput district. The core villages selected for study are given in Table 10.

Table 10: Villages Selected for Study

Village	Community	Block	District
Kondh Pungar	Kondh	Semiliguda	Koraput
Chikalmari	Kondh	Pottangi	Koraput
Putisil	Kondh	Semiliguda	Koraput
Bondaguda	Kondh	Semiliguda	Koraput
Poraja Pungar	Poraja	Semiliguda	Koraput
Gullel	Poraja	Semiliguda	Koraput
Panasput	Poraja	Semiliguda	Koraput
Khojuriguda	Poraja	Semiliguda	Koraput
Chompakanda	Gadaba	Pottangi	Koraput
Dokriguda	Gadaba	Pottangi	Koraput
Billaput	Gadaba	Semiliguda	Koraput
Pipalguda	Gadaba	Semiliguda	Koraput
Sillaiguda	Bonda	Khairput	Malkangiri
Bondapoda	Bonda	Khairput	Malkangiri
Dumripoda	Bonda	Khairput	Malkangiri
Bodbel	Bonda	Khairput	Malkangiri

Village meetings were organised and the objectives and outcome of the study were explained in detail. Appropriate suggestions/modifications given by the villagers were incorporated. The Prior Informed Consent was obtained from the village councils where we intended to do the study, as a part of ethical commitment. The excellent rapport that IRDWSI had with the communities helped the first author to build a close relationship with the villagers.

Voucher specimens were collected and tagged except from sensitive areas such as sacred groves and certain burial grounds and deposited at the Madras Christian College Herbarium. They were also not collected whenever we were prevented from doing so by the knowledge holders. Such circumstances were often met with in

Bonda hills where the knowledge holders were willing to share the knowledge but not the specimens. Such specimens were later collected from non-restricted areas. Information collected was recorded in field notebooks. The specimens were identified once in three months when they were brought back to the laboratory at Chennai, with the help of the Floras such as Flora of the Madras Presidency (Gamble 1915-36), Flora of Orissa (Saxena & Brahmam 1994) and Flora of Visakhapatnam (Rao & Kumari 2002; Rao & Kumari 2008). Plant names were updated by referring to www.tropicos.org, as well as appropriate revisions and monographs. Author citations were standardised using Brummit and Powell (1992). Herbarium specimens were prepared following standard techniques, as prescribed by Forsberg and Sachet (1965).

Effort was taken to attend all festivals and ceremonies happening in the villages which helped in developing a better understanding of their culture, the role played by plants and also helped in strengthening the relationship with the community. Photographs were taken wherever necessary, excluding funeral ceremonies and curing of psychological illness. Oral permission was obtained from the subjects before taking photographs.

Knowledge Holders Involved in the Study

Contrary to the usual practice of selecting the villages on the basis of knowledge holders, we worked with the knowledgeable persons available in the selected villages. All the knowledge holders were married adults above the age of 25, though younger persons also usually joined us on field trips. The group discussions held in villages also witnessed the active participation of enthusiastic children. The tribal children were also great guides during festivals and celebrations. The names of chief knowledge holders and the villages to which they belong are provided in Table 11. The age indicated is only tentative as none of the knowledge holders have records of their age or have kept track of their age. In Bondaguda and Pipalguda, we always worked with a group that had floating members and hence the names are not provided.

Field Study

Throughout the course of fieldwork, the first author was based at Semiliguda from where it was possible for him to stay in constant touch with the communities.

Table 11: Details of Knowledge Holders Involved in the Study

Name of the Knowledge Holder	Gender	Age	Village
Hori Hontal	Male	60	Kondh Pungar
Singru Hontal	Male	30	Kondh Pungar
Sonai Pangi	Female	40	Chikalmari
Kora Mali	Male	50	Putisil
Dona Jani	Male	40	Putisil
Late Jogu Jani	Male	40	Poraja Pungar
Goburdan Kirsani	Male	40	Gullel
Dayanidhi Pujari	Male	38	Gullel
Roghu Jani	Male	50	Panasput
Jamuna Pangi	Female	60	Panasput
Bondu Doli	Male	70	Khojuriguda
Dolo Jimia	Male	65	Khojuriguda
Domu	Male	65	Pakhajola
Sukra Disari	Male	65	Chompakanda
Domu Kalia Nayak	Male	70	Chomapakanda
Gasi Maiji	Male	35	Dokriguda
Sonia Khora	Male	40	Dokriguda
Neelum Maiji	Male	28	Dokriguda
Nondo Toklia	Male	45	Mojikundali
Guru Chepadi	Male	45	Mojikundali
Hontal Guru	Male	95	Billaput
Hadu Hontal	Male	45	Billaput
Suma Bodnaik	Male	45	Bondapoda
Suma Dora	Male	45	Bondapoda
Soma Sisa	Male	45	Silaiguda
Mongla Muduli	Male	50	Dumripoda
Sonia Bodnaik	Male	45	Bodbel

Field trips were conducted in consultation with the resource persons and IRDWSI. A total number of 55 field trips were undertaken. Besides, there were also a number of observational visits undertaken to the villages.

Collection of Information

While preparing the basic framework for the study, manuals such as *Ethnobotany: A People and Plants Conservation Manual* (Martin 1995), *Ethnobotany: Principles and Applications* (Cotton 1997) and *A Handbook of Ethnobotany* (Jain and Mudgal 1999) were consulted in detail. Information was collected by conducting semi-formal interviews. Plant-specific information was collected using the following methods:

(i) Knowledge holders were requested to accompany us to the field and identify the plants used. The information was recorded in the field notes.

(ii) Specimens were brought to the villages and shown to knowledge holders for sharing of information.

(iii) Photographs taken from the forest areas in the vicinity were also used as stimulus.

(iv) Suitable Participatory Rural Appraisal (PRA) techniques were followed in eliciting information from the community for a number of community resources and knowledge-related issues.

Casual chats and passive observation were the chief methods employed to collect data. The interviews were of semi-structured types where no formal questionnaire was employed, but a framework was prepared on the basis of which interviews were held. However, while recording statistical data such as NWFP and agricultural outputs, a structured questionnaire was adopted. The framework for the interviews was based on the following focal points:

HISTORY

— Of the locality
— Of the community
— Of the village
— Of the clan/families

CULTURE

- Culture in current form
- Ancient culture
- Festivals and religion
- Marriages and feasts
- Markets and their role
- Cultural changes
- Factors causing changes
- Impact of changes

SOCIAL SET-UP

- Traditional leadership
- Stratification in the community
- Roles assigned to various clans and the associated practices: past and present

NATURAL RESOURCES

- Types of natural resources
- Present scenario of natural resources
- Natural resources available before
- Assessment of erosion/improvement in natural resources availability
- Conservation of natural resources including forests
- Benefits derived from forests
- Meeting medicinal requirements
    - Custodians of knowledge
    - Imbibing of knowledge and transmission
    - Philosophies associated with the practice
    - Past, present and projected future of the system

AGRICULTURE

- Types of agriculture practised
- Shifting cultivation: past and present
- Traditional and modern cultivars used
- Traditional knowledge involved
- Input and output from traditional cultivars
- Challenges faced
- Agencies promoting change

3

Enumeration of Ethnobotanical Plants

THE study deals with 210 species and two sub-species, belonging to 77 families. Fabaceae is the largest represented family with 16 taxa, immediately followed by poaceae with 14 taxa and Euphorbiaceae with 10 taxa. There are 40 families represented by single taxon. The parts used for various taxa were charted. Of the various plant parts used, fruits and seeds topped the list as they are used from 79 taxa. Whole plant (21 taxa), roots (58 taxa), shoot (4 taxa), stems (23 taxa), bark (37 taxa), leaves (56 taxa) and flowers (19 taxa) are the other parts used (*fig*. 1).

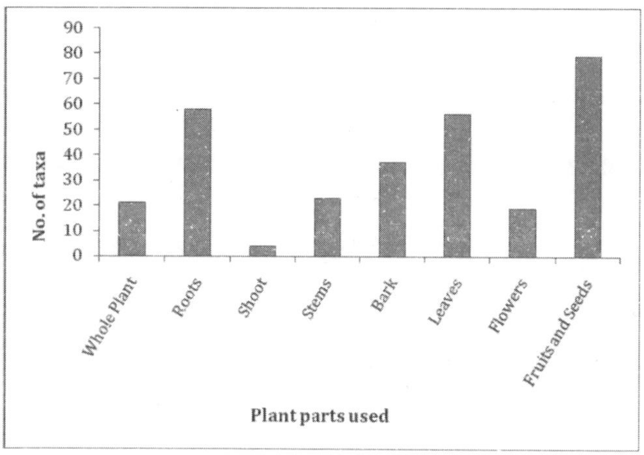

fig. 1: Plant parts used by the communities.

The enumeration follows the following plan.

For each taxon, the correct botanical name is provided in bold letters and the family is indicated in parenthesis. All local names are italicised. Local names are followed by alphabets in parenthesis, which indicates the community/region from which the usage of the name is recorded.

Kuvi local names are indicated by the letter **K**

Poraja local names are indicated by the letter **P**

Gadaba local names are indicated by the letter **G**

Remo (Bonda) local names are indicated by the letter **B**

Names that are regional in nature, common to many communities in the Koraput region, Eastern Ghats or the state of Odisha are denoted by the letter **R**. The names that are used across the nation and also used by the communities are referred to as pan-Indian names, indicated by the letters **PI**. The enumeration deals with both the knowledge in the community/regional domain as well as knowledge in the possession of individuals. Wherever the knowledge is specific to a particular individual, the custodian's name is given in parenthesis, following the use. Since the knowledge is found to vary from community to community for a number of plant resources, the uses are given from a community perspective.

1. Abelmoschus angulosus Wight & Arn. (Malvaceae)

Local name : *Kukura datiki* (K)

Parts used : Bark and seeds

Kondh : (i) Seeds are ground with water and the paste is applied on forehead as febrifuge (Sonai Pangi)

(ii) Bark is peeled and used as cordage

Voucher specimen : FMF 646.

2. Abutilon indicum L. (Malvaceae)

Local name : *Tigercani* (P)

Parts used : Fruits

Poraja : Fruits are dipped in white mud and used as a mould to make designs on the body during a ritual called *Bottu vivag* (*Bottu*=dots, *vivag*=wedding). The ritual is performed by a childless couple as well as couple whose children did not survive for a longer period. Practice of making such designs for play is also common among Poraja girls

Voucher specimen	:		FMF 1131.

3. **Acacia sinuata** (Lour.) Merr. (Mimosaceae)

Local name	:		*Silli, Chilli* (R)
Parts used	:		Stems, leaves and fruits
All Communities	:		Stems are used as firewood; Fruits are sold as well as used as shampoo
Kondh	:		Tender leaves are consumed as greens
Poraja	:	(i)	Fruits are powdered, soaked in water overnight and applied on head while bathing to cure chronic headache
		(ii)	Young leaves are consumed for curing malaria (Dayanidhi Poojari)
Voucher specimen	:		FMF 397.

4. **Achyranthes aspera** L. (Amaranthaceae)

Local name	:		*Rucha bandru* (K), *Ginisher* (B)
Parts used	:		Whole plant, roots and stems
All communities	:		Twigs are used to induce abortion
Kondh and Bonda	:		Root paste is given to pregnant women during labour for safe and easy childbirth.
Bonda	:		Whole plant is ground and applied externally for body pain
Voucher specimen	:		FMF 801.

5. **Acorus calamus** L. (Acoraceae)

Local name	:		*Vasa, Bacha, Vacha* (PI)
Parts used	:		Roots
Poraja	:	(i)	Roots are tied on biceps as talisman to ward off spirits
		(ii)	Roots are also used in domestic rituals
Kondh	:		Roots are chewed and the extract is swallowed for stomach disorders

Voucher specimen : FMF 256.

6. **Adiantum philippense** L. (Adiantaceae) (Fern)

Local name : No name given

Parts used : Whole plant

Bonda : Plants are used in post-delivery rituals. Three bundles of the plants are made, each a fistful. One bundle is given to the mother, one is placed on the roof just above the entrance of the house and one is placed inside the house, beneath the roof. This is believed to ensure long and healthy life of the mother and the infant

Voucher specimen : FMF 797.

7. **Aegle marmelos** (L.) Correa (Rutaceae)

Local name : *Gadhia* (G), *Am pothro* (K), *Bhel* (PI)

Parts used : Stems, leaves and fruits

Gadaba : (i) Leaves are used for worship

(ii) Leaves are fumigated and the smoke is inhaled for curing chickenpox

(iii) Timber is used for house construction

Poraja : (i) Dried leaves are fumigated close to cattle to cure skin infections (Jogu Jani)

(ii) Fruits are squashed with banana and mixed with milk and sugar and consumed as well as offered to God during festivals

Kondh : (i) Leaves are used in domestic rituals to ward off evil spirits

(ii) Ripe fruits are eaten to cure cough

Bonda : Leaves are burnt along with incense and the smoke is inhaled for curing chickenpox

Voucher specimen : FMF 502.

8. **Agave americana** L. (Agavaceae)

Local name	:	*Murga* (R)
Parts used	:	Inflorescence
Poraja	:	Inflorescence peduncles are given to the shepherds as sceptre during annual rituals held during the month of *Poos* (January) that marks the symbolic handing over of shepherdhood
Voucher specimen	:	FMF 1133.

9. **Albizia lebbeck** (L.) Benth. (Mimosaceae)

Local name	:	*Burang* (B)
Parts used	:	Bark
Bonda	:	Powdered bark is used as a fish stupefying agent
Voucher specimen	:	FMF 531.

10. **Aloe vera (L.) Burm. f. (Aloaceae)**

Local name	:	*Hitchu litha* (K), *Gaekumari* (R)
Parts used	:	Leaves
Kondh	:	Leaf extract is applied externally for burns
Voucher specimen	:	FMF 327.

11. **Alstonia scholaris** (L.) R. Br. (Apocynaceae)

Local name	:	*Tala* (K)
Parts used	:	Flowers
Kondh	:	Flowers are consumed along with those of *Mochkund* (*Pterospermum acerifolium*) for dysentery (Hori Hontal)
Voucher specimen	:	FMF 644.

12. **Alstonia venenata** R. Br. (Apocynaceae)

Local name	:	*Mangla jatti* (K), *Koodoi* (P), *Bod Koodoi* (G), *Opudeysaer, Uida* (B)
Parts used	:	Bark, flowers and fruits

Kondh, Gadaba & Poraja	:	Bark is added to *Pendham* along with *Chitramool* (*Plumbago zeylanica*) bark before brewing to increase the potency of the brew
Poraja	:	Bark decoction is consumed for stomach ache (Bondu Doli).
Bonda	:	(i) Fruit paste is consumed to commit suicide
		(ii) Flowers are collected on Sundays without informing anybody and kept above earlobes to ward off head lice. Complete silence has to be maintained from collection to use
		(iii) Bark is consumed to remove 'evil stones' from the body. The symptoms and locality of the pain suggests that the medicine is administered against abdomen pain, probably kidney ailments.
Voucher specimen	:	FMF 541 (*fig*. 17).

13. **Amaranthus spinosus** L. (Amaranthaceae)

Local name	:	*Konta sago* (R)
Parts used	:	Roots and leaves
All communities	:	Leaves are consumed as greens
Kondh	:	Root paste is applied on boils
Voucher specimen	:	FMF 1134.

14. **Ampelocissus tomentosa** (Roth) Planch. (Vitaceae)

Local name	:	*Kendrung puseida* (B)
Parts used	:	Whole plant
Bonda	:	Plant extract is blown into the eyes of humans and cattle for eye ailments (mongla muduli)
Voucher specimen	:	FMF 791.

15. **Andrographis paniculata** Nees (Acanthaceae)

Local name	:	*Aesing* (B), *Booi limb* (R)
Parts used	:	Leaves and flowers

Poraja : (i) Leaves are consumed for headache especially resulting due to loud noises and stress (Dolo Jimia)

(ii) Leaves are consumed for stomach problems

(iii) Leaves are consumed to clear throat

Bonda : Flower extract is given to young children to ward off evil eyes

Voucher specimen : FMF 664.

16. Anogeissus latifolia Wall. (Combretaceae)

Local name : *Thouda* (G)

Parts used : Wood and bark

Gadaba : (i) Wood is used to make handles for plough

(ii) Bark is ground and the paste is administered along with country liquor for cough

(iii) Cocoons of the larva infecting the leaves are sold as NTFP at the rate of Rs. 10/kg

Voucher specimen : FMF 758.

17. Ardisia solanacea Roxb. (Myrsinaceae)

Local name : *Lidi kutty, Reediki* (K), *Goli* (P), *Kitti gocho* (G), *Nbong thithing* (B)

Parts used : Roots, leaves, fruits and seeds

Kondh : (i) Fruits are edible

(ii) Roots are consumed for curing fits

(iii) Leaves are squashed and applied on forehead for headache

Poraja : Two fruits a day are swallowed without chewing for eye pain until cure (bondu doli)

Gadaba : (i) Leaves are used as a substitute for tamarind

(ii) Seeds are roasted and eaten

(iii) Fruits are consumed for relieving eye pain

Bonda	:	Leaves and stems are used in magic rituals along with *Suso kooni* (*Paederia foetida*)
Voucher specimen	:	FMF 1135 (*fig*. 18).

18. Argemone mexicana L. (Papaveraceae)

Local name	:	*Dragu* (K), *Kapura* (B), *Dutra* (R)
Parts used	:	Flowers and seeds
Kondh	:	(i) Seed oil is applied externally for skin diseases in children
		(ii) Flowers are used in domestic rituals
Gadaba	:	Flowers are used in rituals (refer: *Ricinus communis*)
Bonda	:	(i) Seeds are ground and used as styptic in any type of injury (Soma Sisha)
		(ii) Roots are tied as talisman on neck to ward off pain due to fever
		(iii) Prickly stem is used to beat evil-infected persons thrice to drive off evil spirits
Voucher specimen	:	FMF 371.

19. Argyreia nervosa (Burm. f.) Bojer (Convolvulaceae)

Local name	:	*Bod Kasoor, Sathkuari, Oso maal* (R)
Parts used	:	Roots, leaves and seeds
Poraja	:	One seed per day is eaten for three consecutive days to cure indigestion and constipation (Bondu Doli)
Gadaba	:	Leaves are placed on the earlobes of women to facilitate detaching of placenta after delivery
Kondh	:	Roots are ground and the paste is applied to reduce swellings
Voucher specimen	:	FMF 663 (*fig* 19).

20. Aristida setacea Retz. (Poaceae)

Local name	:	*Goot batni* (R)
Parts used	:	Roots and inflorescence
Gadaba	:	(i) Roots are used as an antidote for snakebite (Sukra Dishari)
		(ii) Bundle of inflorescence is used as broom
Voucher specimen	:	FMF 589.

21. Artemisia japonica Thunb. (Asteraceae)

Local name	:	*Dona phul* (P), *Banchi tulsi* (K)
Parts used	:	Roots and leaves
Poraja	:	Leaves are ground and applied on sprains
Kondh	:	Roots are ground with those of *siddi* (*Pedilanthus tithymaloides*) and *Rucha bandru* (*Achyranthes aspera*), tied in a cloth and kept on the pregnant woman's forehead for safe delivery (sonai pangi)
Voucher specimen	:	FMF 331.

22. Artocarpus heterophyllus Lam. (Moraceae)

Local name	:	*Unquising* (B), *Panas* (R)
Parts used	:	Wood and fruits
All communities	:	Fruits are edible; wood is used for house construction
Bonda	:	Liquor is distilled from fruits
Kondh	:	Fruits were traditionally bartered for salt. This has become a rare practice in recent times
Voucher specimen	:	FMF 403.

23. Arundo donax L. (Poaceae)

Local name	:	*Toila bamso* (R)
Parts used	:	Stems

All communities	:	Stem is used for making traditional flutes. Poraja are considered as specialists in making flutes
Voucher specimen	:	FMF 1136.

24. Asparagus racemosus Willd. (Asparagaceae)

Local name	:	*Koori butta* (G), *Isengo* (B), *Pennu heppori* (K), *Dev baatni* (R)
Parts used	:	Tubers, shoots and leaves
Poraja and Bonda	:	Tubers are consumed for stomach ache
Poraja	:	Shoots are tied over the body during harvest of honey to prevent attack of honey bees
Gadaba	:	Leaves and tubers are used in magic rituals to ward off evil spirits
Bonda	:	Shoots are used in magic rituals to drive away evil spirits
Kondh	:	Tubers are eaten for fever accompanied by body pain (Dona Jani)
Voucher specimen	:	FMF 272.

25. Azadirachta indica A. Juss. (Meliaceae)

Local name	:	*Limba* (G), *Limb* (R)
Parts used	:	Leaves
Gadaba	:	Leaves are ground along with *Eda adicher* (*Sida cordata*) and applied for skin diseases especially of private parts (Sukra Dishari)
Poraja	:	Leaves are ground and applied on cattle wounds to expel worms
Voucher specimen	:	FMF 1137.

26. Bambusa bambos (L.) Voss (Poaceae)

Local name	:	*Bamso* (R)
Parts used	:	Stems
All communities	:	Stems are used in making trays, fences, mats and *thonthor* (fish trap)

Voucher specimen : FMF 1138.

27. **Barleria strigosa** Willd. (Acanthaceae)

Local name : *Bathmandi* (R)

Pasts used : Roots

Gadaba : Root extract is drunk for persistent cough

Voucher specimen : FMF 595.

28. **Bauhinia purpurea** L. (Caesalpiniaceae)

Local name : *Koilari* (R)

Parts used : Leaves

All communities : Tender leaves are cooked and consumed as vegetable

Voucher specimen : FMF 1139.

29. **Bauhinia retusa** Roxb. (Caesalpiniaceae)

Local name : *Dondra* (R)

Parts used : Wood and seeds

All communities : Considered as a good timber for house construction

Bonda : Seeds are cooked and eaten. Seeds are said to cause headache if consumed in excess

Voucher specimen : FMF 1140.

30. **Bauhinia vahlii** Wight & Arn. (Caesalpiniaceae)

Local name : *Paaru mada* (K), *Lamu* (B), *Sealli* (R)

Parts used : Stems, bark, leaves and seeds

All communities : Bark is used as cordage; leaves are used to make plates and seeds are edible

Gadaba : (i) Seeds are eaten with *pendham* (rice beer). Seed consumption begins with the festival of *Poos porv*

(ii) Seeds that were stored for three years are soaked in water for three days with salt. The soaked

			seeds are then given for diarrhoea (Sukra Dishari)
Kondh	:	(i)	Leaf extract is poured into nostrils to expel small insects such as caterpillars that might have entered accidentally (Dona Jani)
		(ii)	It is said that the ropes made of *Paaru maada* were used to wrap the bodies of the dead in traditional cremations
Voucher specimen	:		FMF 365 (*fig*. 20).

31. Bauhinia variegata L. (Caesalpiniaceae)

Local name	:		*Kosanada, Kochanada* (R)
Parts used	:		Wood, flowers and seeds
Gadaba	:	(i)	Seeds and flowers are roasted and eaten as well as used to make curry
		(ii)	Timber is used in construction and for making plough
Voucher specimen	:		FMF 1141 (*fig*. 21).

32. Bidens pilosa L. (Asteraceae)

Local name	:	*Sukhali butta* (R)
Parts used	:	Inflorescence
Gadaba	:	Inflorescence is ground with jaggery, *Lach kutti* (*Mimosa pudica*) and given to the opposite sex, especially girls without their knowledge to make her amorous (Sukra Dishari).
Voucher specimen	:	FMF 363.

33. Bixa orellana L. (Bixaceae)

Local name	:	*Jaapra* (PI)
Parts used	:	Seeds
All communities	:	Seeds are sold at Rs. 13-15/kg when there is a market demand
Kondh	:	Seeds are used to colour borders of white sari worn during festivals

Voucher specimen	:		FMF 13.

34. Boerhavia diffusa L. (Nyctaginaceae)

Local name	:		*Attuva kutcha* (K)
Parts used	:		Roots
Kondh	:		Roots are eaten raw to cure diarrhoea
Voucher specimen	:		FMF 334.

35. Bombax ceiba L. (Bombacaceae)

Local name	:		*Lekha marnu* (K), *Simili* (R)
Parts used	:		Wood and bark
Gadaba	:		Wood is used for making wheel of the miniature cart for the goddess during festivals
Poraja	:	(i)	Trunk is used as poles
		(ii)	Bark paste is applied on boils or swellings that protrude to the size of a fist from any part of the body to suppress the protrusion (Dolo Jimia)
Kondh	:	(i)	Bark is crushed and used as "soap" for body wash
		(ii)	This tree is very much revered in the Kondh folklore. The flowers are considered as an important element of spring and find place in folk songs, especially romantic ones. It is also said that the ancestor couple of the Kondh lived in a boat carved out of *simili* tree
Voucher specimen	:		FMF 662 (*fig.* 22).

36. Brassica oleracea L. (Brassicaceae)

Local name	:		*Bonda gopi* (R)
Parts used	:		Leaves
All communities	:		Cultivated as a market crop, except Bonda. Seeds for sowing are obtained from the markets
Voucher specimen	:		FMF 1142.

37. Breynia retusa (Dennst.) Alston (Euphorbiaceae)

Local name	:	*Koyunatheri* (K), *Muangsebo, Simosing, Phulbani* (B), *Sathodi gosch, Hathi Girli* (R)
Parts used	:	Roots, stem, bark, leaves and flowers
Gadaba	:	Root and bark are prescribed for dysentery (Sukra Dishari)
Poraja	:	Mother is advised to brush teeth with the twig for three consecutive days if her infant's stool happens to be green (Raghu Jani)
Bonda	: (i)	Twigs are used as toothbrush to cure toothache
	(ii)	Flowers are consumed along with root extract for dysentery (Mongla Muduli)
	(iii)	Leaves are placed above the earlobes to ease child delivery
	(iv)	Roots are ground and eaten in empty stomach for dysentery, twice a day for two days
Kondh	:	Root is administered for dysentery and fits (Dona Jani)
Voucher specimen	:	FMF 461 (*fig.* 23).

38. Bridelia retusa Spreng. (Euphorbiaceae)

Local name	:	*Riduwooti* (K), *Kosi* (R)
Parts used	:	Branches
Gadaba	:	Tree is considered sacred and the branches are offered to gods during rituals
Voucher specimen	:	FMF 270.

39. Bryophyllum calycinum Salisb. (Crassulaceae)

Local name	:	*Pudubullu* (B), *Pothru kaja* (K)
Parts used	:	Roots and leaves
Bonda	:	Leaves and roots are ground and the paste is administered for dysentery in goats
Kondh	:	Three leaves with 21 pepper seeds are chewed

			and eaten like *pan* once for watery diarrhoea (Hori Hontal)
Voucher specimen	:		FMF 520.

40. Buchanania lanzan Spreng. (Anacardiaceae)

Local name	:		*Jarri* (P), *Chargulul* (G), *Charcoli* (R)
Parts used	:		Roots, wood, bark and fruits
All communities	:	(i)	Fruits are edible (*Note*: unlike the other regions, tribal communities in the Koraput region eat the pericarp and not the kernel)
		(ii)	Wood is used as firewood, to make agricultural implements and for house construction
Poraja	:	(i)	Bark and roots are made into a paste with bark of *Sindhi* (*Phoenix acaulis*) and flour of *Suan* (*Panicum miliaceum*), roasted and made into *uppuma* (traditional dish) and eaten by new mothers when they feel pain in the navel region (Raghu Jani)
		(ii)	Bark is mixed with that of *Kumi* (*Gardenia latifolia*), *Noondri* (*Syzygium cumini*), *Saaj* (*Terminalia tomentosa*), coconut gratings, ground using stone and the extract is consumed for dysentery (Raghu Jani)
Voucher specimen	:		FMF 458 (*fig*. 24).

41. Butea monosperma (Lam.) Taub. (Fabaceae)

Local name	:	*Buduli* (R)
Parts used	:	Leaves
All communities	:	Leaves are used to make plates. Rarely used owing to scarce occurrence
Voucher specimen	:	FMF 1143.

42. Caesalpinia decapetala (Roth) Alston (Caesalpiniaceae)

Local name	:	*Chillor* (R)
Parts used	:	Seeds

Kondh	:	Plants are planted as hedge. Seeds are sold at Rs.20-21/kg in the local market (*Note*: The plants were noticed in only one village — Kondh Pungar, said to be introduced to the village by the forest department as a hedge plant)
Voucher specimen	:	FMF 731.

43. Cajanus cajan (L.) Huth (Fabaceae)

Local name	:	*Kandul* (R)
Parts used	:	Seeds
All communities	:	Cultivated for its pulses
Voucher specimen	:	FMF 1144.

44. Calotropis gigantea (L.) W.T. Aiton (Asclepiadaceae)

Local name	:	*Arak* (G), *Thua* (P)
Parts used	:	Latex
Gadaba	:	(i) Latex is applied on the finger nails for eye problems (Sukra Dishari)
		(ii) Latex is also applied for minor wounds (Sukra Dishari). *Note:* They also identify it as "poisonous".
Poraja	:	Latex along with that of *Takamal* (*Plumeria rubra*), sap of *Juljuli* (*Crotalaria pallida*), three types of any edible oil and flour of three-year-old finger millets are ground, made into two tablets and given to women for a day in the morning and evening to improve health and immunity
Voucher specimen	:	FMF 350.

45. Cannabis sativa L. (Cannabaceae)

Local name	:	*Ganjai* (PI)
Parts used	:	Leaves
Poraja	:	Leaves are powdered and consumed along with

		that of *Thoska* (*Terminalia chebula*) for cough (Dayanidhi Pujari)
Voucher specimen	:	FMF 1145.

46. Canscora diffusa (Vahl) R. Br. ex Roem. & Schult. (Gentianaceae)

Local name	:	*Goising koi* (B)
Parts used	:	Whole plant
Bonda	:	Whole plant is ground, mixed with paddy husk and given to fowls that are extremely ill (Suma Bodnaik, Suma Dora)
Voucher specimen	:	FMF 702.

47. Canthium dicoccum Merr. (Rubiaceae)

Local name	:	*Maettera* (K), *Olang ser* (B), *Ani mandini* (R)
Parts used	:	Wood and leaves
Poraja	:	Considered to be a strong termite-proof tree and hence preferred for house construction
Bonda	:	Leaf is coated with mustard oil, heated over the flame and rubbed externally for body pain (Soma Sisa)
Voucher specimen	:	FMF 470.

48. Capparis olacifolia Hook.f. & Thomson (Capparaceae)

Local name	:	No name given
Parts used	:	Roots
Kondh	:	Root bark is eaten raw for stomach pain (Hori Hontal)
Voucher specimen	:	FMF 723.

49. Capsicum annuum L. (Solanaceae)

Local name	:	*Morcho, Morich* (R)
Parts used	:	Fruits
All communities	:	Cultivated for its fruits. A favourite crop of the

		Bonda. Seed materials are freely exchanged in the local markets
Voucher specimen	:	FMF 1146.

50. Carica papaya L. (Caricaceae)

Local name	:	*Amruth* (R)
Parts used	:	Fruits
Poraja	:	Papaya fruits are ground along with tender leaves of *Amar* (*Pedilanthus tithymaloides*), *Doodh mal* (*Cryptolepis buchananii*), *Laden* (long black Earthworms), *Jha* (earthworm) and eaten as galactagogue (Raghu Jani)
Gadaba	:	Papaya fruits are ground with roots of *Doodh mal* (*Cryptolepis buchananii*), *Sindhi* (*Phoenix acaulis*) and eaten as galactagogue
Voucher specimen	:	FMF 479.

51. Caryota urens L. (Arecaceae)

Local name	:	*Marakadu* (K), *Thad marin* (G), *Solap, Salop* (R)
Parts used	:	Apical meristems, leaves and inflorescence
All communities	:	(i) Inflorescence is tapped and the sap consumed as such or brewed to obtain toddy. Sometimes, it is distilled to obtain liquor
		(ii) Apical meristem of dead tree is eaten
		(iii) Leaves are used in decoration during festivals and marriages
Bonda	:	The tree is considered to be a priceless possession by the Bonda. The Bonda have a unique ownership pattern over the tree that the tree belongs to the person who had planted (protected, in most cases) it, regardless of the ownership of the land. The Bonda go in groups to the tree, sit beneath it in circles and then the gourd containing the toddy is passed from one person to another while each person takes a mouthful of toddy from it

Voucher specimen	:		FMF 402 (*fig*. 25).

52. **Casearia graveolens** Dalzell (Flacourtiaceae)

Local name	:		*Kirchi* (G), *Girchi* (P), *Burey* (B), *Kissi* (K)
Parts used	:		Wood, flowers and seeds
Gadaba and Poraja	:		Seeds are used as a source of non-edible oil
Kondh and Gadaba	:		Wood is used in making furniture and for house construction
Bonda	:	(i)	Seeds are used as a source of edible oil
		(ii)	Flowers are used as vegetable in curry
Voucher specimen	:		FMF 466 (*fig*. 26).

53. **Cassia fistula** L. (Caesalpiniaceae)

Local name	:		*Boondeni* (K), *Sunali* (B), *Suneri* (R)
Parts used	:		Bark, stem, flowers and fruits
Poraja	:	(i)	Bark is powdered and consumed as febrifuge
		(ii)	Flowers and bark are eaten for prolonged illness of any kind, to boost immunity (Raghu Jani)
		(iii)	Twigs with coconut are used in rituals to cure chronic illness
Kondh	:		Wood is used to make rice pounder
Gadaba	:		Branches with the leaves are erected at the four corners of the agricultural field to prevent evil spirits from entering
Bonda	:		Fruits are kept along with banana fruits while storing to facilitate early ripening of banana fruits
Voucher specimen	:		FMF 472 (*fig*. 27).

54. **Celosia argentea** L. (Amaranthaceae)

Local name	:		*Gungi* (K), *Buiaur* (B)
Parts used	:		Leaves

All communities	:		Leaves are used as greens. It is said that the plant is also one of the famine foods. Bonda is the only community that frequently consumes this now. Others rarely consume it
Voucher specimen	:		FMF 1147.

55. Centella asiatica (L.) Urb. (Apiaceae)

Local name	:		*Duakusey* (G)
Parts used	:		Leaves
All communities	:		Leaves are consumed as greens
Gadaba	:		Leaves are cooked and eaten for stomach ache
Voucher specimen	:		FMF 1148.

56. Chionanthus mala-elengi (Dennst.) P.S. Green (Oleaceae)

Local name	:		*Odi jam* (K), *Katki* (G), *Adu jam* (P, G)
Parts used	:		Wood, bark and fruits
All communities	:	(i)	Timber is used to make furniture and plough handle
		(ii)	Fruits are said to be eaten by bears
Poraja	:	(i)	Bark paste is eaten for stomach pain (Raghu Jani)
		(ii)	Bark powder is mixed with the toddy of *Caryota urens* during the process of tapping to increase its intoxication potency
Voucher specimen	:		FMF 647.

57. Cinnamomum caudatum Nees (Lauraceae)

Local name	:		*Rupini* (K)
Parts used	:		Wood and bark
Kondh	:	(i)	Wood is used in house construction
		(ii)	Bark paste is applied on the forehead of children for fever (Sonai Pangi)
Voucher specimen	:		FMF 784 (*fig*. 28).

58. Cissampelos pareira L. (Menispermaceae)

Local name : *Dindeni mada* (K), *Praam* (B), *Betti cher, Beetti* (P)

Parts used : Roots and seeds

Poraja : (i) Root paste is consumed for curing fits (Bondu Doli)

(ii) Root paste is consumed for stomach ache (Bondu Doli)

(iii) Seeds are ground, mixed with *Karanch* (*Pongamia pinnata*) oil and applied for scabies (Jogu Jani)

Bonda : Roots are ground with that of *Siripupa* (*Pogostemon benghalensis*) and administered for stomach ache resulting from diarrhoea (Mongla Muduli)

Kondh : Root extract is mixed with extracts of whole plant of *gori saveli* (*Mimosa pudica*) and roots of *Poosi heeru* (*Rauvolfia serpentina*) and consumed when pain is felt in the abdomen resembling that of an "electric shock". Though the herbalist does not term it as urinary problem, the fact that the plant derives its name from the urinary bladder (*dindenni*) suggests so (Hori Hontal)

Voucher specimen : FMF 616.

59. Clausena excavata Burm. f. (Rutaceae)

Local name : *Kaja kunuva* (K)

Parts used : Leaves

Kondh : Leaves are dried, ground and applied for eye problems of cattle (Shirma Dishari)

Voucher specimen : FMF 782.

60. Clematis gouriana Roxb. (Ranunculaceae)

Local name : *Gana madi* (R)

Parts used	:	Leaves
All communities	:	Plant is mistakenly used as substitute for *Clematis roylei* at times due to confused identity
Voucher specimen	:	FMF 1149.

61. Clematis roylei Rehder (Ranunculaceae)

Local name	:	*Gana madi* (R), *Thamisae* (B)
Parts used	:	Whole plant and tender leaves
All communities	:	Tender leaves are squashed and the aroma is inhaled for cough and congestion
Bonda	:	Plants are used in domestic rituals to chase evil spirits
Voucher specimen	:	FMF 536 (*fig*. 29).

62. Clerodendrum serratum Spreng. (Lamiaceae)

Local name	:	*Mahapru phool* (R)
Parts used	:	Bark and inflorescence
Poraja	:	(i) Bark past applied on the spot of snake bite (Raghu Jani)
		(ii) Inflorescence is planted vertically in agricultural fields to ward off evil eyes
Voucher specimen	:	FMF 656.

63. Clerodendrum viscosum Vent. (Lamiaceae)

Local name	:	*Bandari* (R)
Parts used	:	Roots
Poraja	:	Root paste is consumed for stomach ache (Raghu Jani)
Voucher specimen	:	FMF 471.

64. Cocos nucifera L. (Arecaceae)

Local name	:	*Nadia* (R)
Parts used	:	Fruits

Poraja : (i) Coconut gratings are mixed with barks of *kumi* (*Gardenia latifolia*), *jarri* (*Buchanania lanzan*), *noondri* (*Syzygium cumini*), *saaj* (*Terminalia tomentosa*) and ground using stone. The extract is consumed for dysentery

(ii) Coconut, along with twigs of *suneli* (*Cassia fistula*), are used in rituals for prolonged illness

(iii) Coconut shell ash is mixed with coconut oil and applied on burns (Jamuna Pangi)

Voucher specimen : FMF 1150.

65. **Colebrookea oppositifolia** Sm. (Lamiaceae)

Local name : *Kinbang bau* (B)

Parts used : Roots

Bonda : Roots are ground and applied on forehead as febrifuge

Voucher specimen : FMF 1151.

66. **Costus speciosus** Sm. (Costaceae)

Local name : *Gimae sirasiv* (B), *Bommoki* (P), *Kaev kondha* (R)

Parts used : Rhizomes

Bonda : Rhizome is ground with black pepper and eaten for cough (Mongla Muduli)

Kondh : Rhizome paste is eaten for stomach ache

Poraja : Rhizome paste is applied on forehead to keep the head cool and to cure migraine

Gadaba : Rhizome paste is consumed as well as applied on the forehead as febrifuge

Voucher specimen : FMF 813 (*fig.* 30).

67. **Crotalaria pallida** Dryand. (Fabaceae)

Local Name : *Kooisesing* (B), *Juljuli* (R)

Parts used : Roots and stem

Poraja	:	Root paste is consumed for fits (Bondu Doli)
Bonda	:	Stem is used as toothbrush to cure toothache
Voucher specimen	:	FMF 689.

68. **Cryptolepis buchananii** Roem. & Schult. (Asclepiadaceae)

Local name	:	*Paalu mada* (K), *Doodh mal* (P), *Doodh oso* (G), *Ae_ooi* (B)
Parts used	:	Roots and leaves
Kondh	:	(i) Roots and leaves are given to cattle to improve lactation
		(ii) Root alone is eaten as galactagogue by nursing mothers
Poraja	:	Tender leaves are ground with that of *amar* (*Pedilanthus tithymaloides*), Papaya fruits, *laden* (long black earthworms) and *jha* (earthworm) and eaten as galactagogue (Raghu Jani)
Gadaba	:	(i) Roots of *Cryptolepis* are ground with that of *sindhi* (*Phoenix acaulis*), papaya fruits and eaten as galactagogue
		(ii) The Gadaba have deep emotional relationship with this plant. Plant is seen engraved on both the sides of doors, even in modern houses. Engraving is done often in a stylised form bearing flowers of *Chrysanthemum*.
Bonda	:	Leaves are crushed and eaten to ward off evil spirits
Voucher specimen	:	FMF 318 (*fig*. 31).

69. **Cucumis sativus** L. (Cucurbitaceae)

Local name	:	*Kakodi* (R)
Parts used	:	Fruits
All communities	:	The plant is cultivated for its fruits that are consumed as vegetable
Voucher specimen	:	FMF 1152.

70. Curculigo orchioides Gaertn. (Hypoxidaceae)

Local name : *Emur thuthur* (B)

Parts used : Roots

Bonda : Roots are ground with mushroom growing on bamboo and the paste is applied on boils (Mongla Muduli)

Voucher specimen : FMF 795.

71. Curculigo trichocarpa (Wight) Bennet & Raizoda (Hypoxidaceae)

Local name : *Emur thuthur* (B)

Parts used : Roots

Bonda : *C. trichocarpa* and *C. orchioides* are considered to be the same species by the Bonda and hence used for the same purpose under the same name

Voucher specimen : FMF 1153.

72. Cymbopogon gidarba (Buch.-Ham. ex Steud.) Haines (Poaceae)

Local name : *Singani piri* (R)

Parts used : Whole plant

All communities : Culms are used for thatching the traditional houses

Voucher specimen : FMF 1154.

73. Cynodon dactylon (L.) Pers. (Poaceae)

Local name : *Saepsa* (B), *Garki* (K), *Thuba gaso* (R)

Parts used : Whole plant and roots

Bonda : Root paste is applied on injuries

Kondh : Whole plant is ground and administered as emetic (Sonai Pangi)

Voucher specimen : FMF 711.

74. Dalbergia ougeinensis Roxb. (Fabaceae)

Local name	:	*Bondan* (R)
Parts used	:	Wood
All communities	:	Timber is used for house construction
Voucher specimen	:	FMF 760 (*fig*. 32).

75. Datura stramonium L. (Solanaceae)

Local name	:	*Thuthura* (K), *Dathura* (PI)
Parts used	:	Roots, latex and seeds
Kondh	:	Leaf latex is applied on the base of hairs along the hairline to arrest hairfall (Dona Jani)
Bonda	:	Roots and seeds are powdered and consumed during night for seven consecutive Fridays for diarrhoea combined with body ache. The dosage has to be strictly followed since the plant is also poisonous (Sonia Bodnaik)
Voucher specimen	:	FMF 560.

76. Debregeasia velutina Gaudich. (Urticaceae)

Local name	:	*Borodo goch* (K)
Parts used	:	Wood and bark
Kondh	:	(i) Wood is used for house construction purposes and is valued for its insect resistant properties
		(ii) Bark paste is applied externally for leg injuries of cattle
		(iii) Kondh of Putisil believe that the plant and *Ricinus communis* have a mutually symbiotic relationship. When the area where *Borodo* tree is available happens to be burnt, *R. communis* appears quickly, provided that the *Borodo* tree is available in the vicinity
Voucher specimen	:	FMF 398.

77. Dendrocalamus strictus Nees (Poaceae)

Local name	:	*Kordi bamso* (R)
Parts used	:	Tender shoots
All communities	:	Tender shoots are soaked in running water for three continuous days or boiled repeatedly several times to reduce the toxicity and then consumed as vegetable
Voucher specimen	:	FMF 1155.

78. Desmodium gangeticum (L.) DC. (Fabaceae)

Local name	:	*Kanna kathiri* (P)
Parts used	:	Roots
Poraja	:	Roots are consumed for curing fits (Bondu Doli)
Voucher specimen	:	FMF 695.

79. Desmodium triquetrum DC. (Fabaceae)

Local name	:	*Baasing* (B)
Parts used	:	Stem
Bonda	:	Stem is tied to the biceps as talisman to ward off attacks from wild animals and spirits
Voucher specimen	:	FMF 794.

80. Dioscorea bulbifera L. (Dioscoreaceae)

Local name	:		*Podhul kurrey* (G), *Pitta kanda* (R)
Parts used	:		Tubers
All communities	:	(i)	Bitter tubers are sliced and boiled repeatedly or sliced and left immersed in running water for a day, cooked and eaten. A less preferred wild food
		(ii)	Tubers are offered to the cattle during the festival of *Poos porv* as a mark of honour
		(iii)	Tubers are consumed for stomach ache. Believed to be effective due to its bitterness

Voucher specimen	:	FMF 281.

81. **Dioscorea glabra** Roxb. (Dioscoreaceae)

Local name	:	*Peet kanda* (R)
Parts used	:	Tubers
All communities	:	A much preferred wild food. Tubers are cooked and relished upon
Voucher specimen	:	FMF 1156.

82. **Dioscorea oppositifolia** L. (Dioscoreaceae)

Local name	:	*Doro kunaya* (K), *Dorgai, Tharga kanda* (R)
Parts used	:	Tubers
All communities	:	Tubers are cooked and eaten
Voucher specimen	:	FMF 243.

83. **Dioscorea pentaphylla** L. (Dioscoreaceae)

Local name	:	*Sorenda kanda* (R)
Parts used	:	Tubers
All communities	:	Tubers are boiled repeatedly and eaten. A less-preferred wild food
Gadaba	:	The yam is used as a sacrificial surrogate for pig, when a pig is not affordable
Voucher specimen	:	FMF 1157.

84. **Diospyros melanoxylon** Roxb. (Ebenaceae)

Local name	:	*Thirae* (B), *Kendu* (PI)
Parts used	:	Fruits
Bonda	:	Fruits are edible.
Voucher specimen	:	FMF 1158.

85. **Diospyros sylvatica** Roxb. (Ebenaceae)

Local name	:	*Potta thoru* (K), *Makorkendu* (G)
Parts used	:	Wood and fruits

Kondh, Gadaba, and Poraja	:	Wood is used for making furniture
Poraja	:	Fruits and wood are used for performing rituals during cremation
Voucher specimen	:	FMF 749.

86. Diplocyclos palmatus (L.) C. Jeffrey (Cucurbitaceae)

Local name	:	*Bana batti* (G), *Giga bulu* (B)
Parts used	:	Roots and fruits
Gadaba	:	(i) Root paste is consumed for nausea due to fever (Sukra Dishari)
		(ii) Roots are ground and applied for joint pains
Bonda	:	Fruits are consumed for stomach ache
Voucher specimen	:	FMF 361.

87. Dracaena terniflora Roxb. (Dracaenaceae)

Local name	:	*Sing sop* (B)
Parts used	:	Leaves
Bonda	:	Leaves are soaked in water for a few hours and then poured over people frightened by tiger along with a chant which crudely means "Tiger tiger go away, do not follow me wherever I go". (What is called as "tiger" is in fact a leopard as tigers are locally extinct)
Voucher specimen	:	FMF 1159.

88. Elephantopus scaber L. (Asteraceae)

Local name	:	*Ondra kusuma, Thath kida* (P), *Mothiri* (B)
Parts used	:	Roots and seeds
Poraja	:	(i) Three seeds are eaten together as a single dose to cure constipation
		(ii) The plant is considered as the male form of *kusuma* (*Eryngium foetidum*).

(iii) Root is chewed with seven crystals of common salt for gum ache, toothache and swellings in the buccal cavity (Dayanidhi Pujari)

Bonda : Root paste is consumed thrice a day for three consecutive days for stomach ache (Sonia Bodnaik)

Voucher specimen : FMF 1160.

89. Eleusine coracana Gaertn. (Poaceae)

Local name : *Mandia* (R)

Parts used : Seeds

All communities : Cultivated as a dryland crop for its grains that form a staple food. Used to make *pendham* and *landha*, two traditional beverages

Voucher specimen : FMF 1161.

90. Eryngium foetidum L. (Apiaceae)

Local name : *Kusuma* (P), *Bono dhania* (K)

Parts used : Roots and fruits

Poraja : Fruits are consumed as laxative

Kondh : Roots are eaten for deworming

Voucher specimen : FMF 735 (b).

91. Euphorbia hirta L. (Euphorbiaceae)

Local Name : *Godhuni laatta* (G), *Doodh mal* (R)

Parts used : Latex

All communities : Plant latex, along with castor oil and carbon from the wick of burnt lamp are used as ink in tattooing. Tattooing is considered to be a speciality of Poraja women

Poraja : Latex is administered externally for eyesore on the corner of the eye closer to the temple region for three consecutive days (Bondu Doli)

Voucher specimen : FMF 608.

92. Euphorbia nivulia Buch.-Ham. (Euphorbiaceae)

Local name	:	*Juvi* (K), *Jilleni* (P), *Tua* (G)
Parts used	:	Latex, bark and leaves
Kondh	:	Latex is applied just above the eye for eye problems of cattle. *Note:* Not applied on the eyelids or eye as it can cause blindness
Poraja	:	Bark with that of *takamal* (*Plumeria rubra*) is ground, burnt, made into a cake (*pitta*) with *mandia* (*Eleusine coracana*) flour and eaten for facial and abdominal swellings
Gadaba	:	Leaves are crushed and put in shallow waters to stupefy fish
Voucher specimen	:	FMF 1162.

93. Ficus arnottiana (Miq.) Miq. (Moraceae)

Local name	:	*Jodi* (R)
Poraja, Gadaba and Bonda	:	Considered as sacred and planted in the *sadhur* — the place where the village council gathers
Voucher specimen	:	FMF 1163.

94. Ficus asperrima Teijsm. & Binn. (Moraceae)

Local name	:	*Gana, Gina* (B), *Karsali* (G), *Grawi/Garwi* (K)
Parts used	:	Roots, latex and leaves
Bonda	:	(i) Roots are tied as talisman on neck for fever
		(ii) Latex is applied on cattle wounds for deworming
Kondh	:	Leaves are used as a natural emery paper
Voucher specimen	:	FMF 745.

95. Ficus benghalensis L. (Moraceae)

Local name	:	*Boro goch* (R)
All communities	:	Worshipped by communities adopting Hinduism

Voucher specimen : FMF 1164.

96. Ficus religiosa L. (Moraceae)

Local name : *Aalmarin* (G), *Peppedi* (K), *Pipal* (PI)

Gadaba : The tree is planted in *sadhur* (village meeting place) as a recent replacement for *jodi* (*Ficus arnottiana*)

Voucher specimen : FMF 839.

97. Ficus semicordata Buch.-Ham. ex Sm. (Moraceae)

Local name : *Drohi* (K), *Podhoi* (P), *Vaertha* (G)

Parts used : Bark and fruits

All communities : Fruits are edible

Poraja : Bark is used to mend bamboo baskets and also used as cordage

Voucher specimen : FMF 754 (*fig.* 33).

98. Firmiana colorata (Roxb.) R. Br. (Sterculiaceae)

Local name : *Goos sothodi* (P), *Sirumool phul* (G), *Tingri*

Parts used : Bark and flowers.

Gadaba : Bark is consumed to ease post-delivery pain.

Poraja : (i) Flowers are used in rituals during the festival of *Choith porv* and also to decorate the canes to which the hunted birds are tied, during Annual Ritual Hunt

(ii) Bark extract is consumed for joint pain (Jogu Jani)

Voucher specimen : FMF 1165 (*fig.* 34).

99. Fleurya interrupta Gaudich. (Urticaceae)

Local name : *Guna gocho* (K), *Kareng* (G)

Parts used : Bark

All communities : Bark is used as minor cordage

Bonda and Gadaba : Bark is used for making threads for weaving clothes. Threads for weaving the traditional dress of Gadaba women (*kareng*) and the lower wear of Bonda women (*ringa*) were obtained from this plant. Modern cotton threads obtained from the markets have replaced these traditional threads

Voucher specimen : FMF 651.

100. Furcraea foetida (L.) Haw. (Agavaceae)

Local name : *Murga* (B), *God kuari* (P)

Parts used : Leaves

Bonda : Tender leaves are burnt and patted on stomach for stomach ache (Soma Sisa)

Poraja : Leaves or roots are kept on the ear to ease delivery. Has to be removed as soon as the complication is over or else may turn fatal

Voucher specimen : FMF 1166 (*fig*. 35).

101. Galactia tenuiflora (Spreng.) Wight & Arn. (Fabaceae)

Local name : *San loho mada* (K)

Parts used : Stem

Kondh : Stems are used as minor cordage.

Voucher specimen : FMF 732.

102. Gardenia latifolia Aiton (Rubiaceae)

Local name : *Kumi* (P), *Bongra* (G)

Parts used : Bark and fruits.

All communities : Children use fruits as "top" and as a trajectile in catapult

Poraja : Bark is ground with that of *jarri* (*Buchanania lanzan*), *noondri* (*Syzygium cumini*) and *saaj* (*Terminalia tomentosa*), coconut gratings and the extract is consumed for dysentery (Raghu Jani)

Voucher specimen	:	FMF 463.

103. Garuga pinnata Roxb. (Burseraceae)

Local Name	:	*Modhoi* (R)
Parts used	:	Branches
Poraja	:	Branch is given to the shepherd as sceptre along with finger millets to ensure good health of the cattle for a year
Gadaba	:	Branch piece erected as stump 15 cm above the ground in a corner as a ceremonial mark of house construction. This stump is also worshipped as family deity as long as the family resides in the house
Voucher specimen	:	FMF 1167.

104. Gloriosa superba L. (Colchicaceae)

Local name	:	*Bag mundi* (R)
Parts used	:	Rhizome
Gadaba	:	Corm paste is applied externally for goitre and other swellings in the neck region
Voucher specimen	:	FMF 1168.

105. Gmelina arborea Roxb. (Lamiaceae)

Local name	:	*Gombhari* (R)
Parts used	:	Wood
Kondh	:	Wood is used for making furniture
Voucher specimen	:	FMF 1169.

106. Grewia tiliifolia Vahl (Tiliaceae)

Local name	:	*Dhamen* (R)
Parts used	:	Wood
All communities	:	Timber is used for house construction
Voucher specimen	:	FMF 493.

107. Guizotia abyssinica (L.f.) Cass. (Asteraceae)

Local name : *Olsi* (R)

Parts used : Seeds

All communities : Cultivated as a commercial oil crop. Oil is also used for culinary purposes

Poraja : Leaves of *tendca* (*Woodfordia fruticosa*) are fried in *olsi* oil, ground into a paste and applied for wounds appearing adjacent to the earlobes

Voucher specimen : FMF 1170.

108. Helicteres isora L. (Sterculiaceae)

Local name : *Poolagala* (B)

Parts used : Bark and fruits

Bonda : (i) Bark is used as cordage

(ii) Fruits when tied to *salop* tree (*Caryota urens*) are believed to increase the potency of the toddy

Voucher specimen : FMF 808.

109. Hemidesmus indicus (L.) R. Br. (Asclepiadaceae)

Local name : *Insorlim* (B), *Chemidimulo, Sugandhi* (R)

Parts used : Roots and leaves

Bonda : Leaf or root paste is consumed for stomach ache

Voucher specimen : FMF 1198.

110. Hemionitis arifolia (Burm. f.) Moore (Hemionitidaceae) (Fern)

Local name : *Kattakar* (P), *Kaat porey* (G)

Parts used : Whole plant and roots

Poraja and Gadaba : Roots are tied as talisman on neck or arm to cure fever in children

Gadaba : (i) Plants are used in rituals performed prior to the erection of a new cowshed

(ii) Plant is tied as talisman on the neck of new-

			born children before exposing them to non-family members
Voucher specimen	:		FMF 1203.

111. Hiptage benghalensis (L.) Kurz (Malpighiaceae)

Local name	:		*Loho mada* (K)
Parts used	:		Stem and bark
All communities	:		Stems are used as cordage
Kondh	:		Bark paste is applied on cuts
Voucher specimen	:		FMF 1204.

112. Holarrhena pubescens (Buch.-Ham.) Wall. ex G. Don (Apocynaceae)

Local name	:		*Paalosi* (P), *Paalsi* (K), *Koodoi* (G)
Parts used	:		Roots, wood and bark.
Poraja and Kondh	:		Wood is used for making spade handle, yoke and for firewood
Kondh	:	(i)	Bark powder is added to *pendham* and *salop* for increasing the potency of the brew
		(ii)	Stem was used to make traditional combs due to its anti-lice property
Gadaba	:		Root paste is administered to women for all kinds of post-delivery complications
Voucher specimen	:		FMF 1201.

113. Hydrocotyle rotundifolia Roxb. (Apiaceae)

Local name	:		*Kadpi* (P)
Parts used	:		Whole plant
Poraja	:		Whole plants are chewed to cure white spots appearing on the tongue (Bondu Doli)
Voucher specimen	:		FMF 611 (*fig.* 36).

114. Hyptis suaveolens (L.) Poit. (Lamiaceae)

Local name	:		*Sachra* (R)

Parts used : Seeds

Poraja : Seeds were smashed and used to wash body in the past

Voucher specimen : FMF 196.

115. Indigofera cassioides Rottler. ex DC. (Fabaceae)

Local name : *Leda* (K), *Girli* (R)

Parts used : Roots, bark and flowers

Kondh : (i) Flowers are roasted with red gram and eaten

(ii) Roots are ground with pepper and the paste is administered for asthma

Poraja : (i) Bark is made into a paste with that of *jarri* (*Buchanania lanzan*), roots of *sindhi* (*Phoenix acaulis*) and flour of *suan* (*Panicum miliaceum*), dry roasted and made into *uppuma* (traditional dish) and eaten when pain is felt in the naval region by new mothers (Raghu Jani)

(ii) Flowers are used as vegetable

Voucher specimen : FMF 464 (*fig*. 37).

116. Justicia adhatoda L. (Acanthaceae)

Local name : *Mohra* (R)

Parts used : Leaves

Poraja : Leaf paste is applied on wounds

Poraja and Kondh : Leaf paste is applied on wounds in cattle

Voucher specimen : FMF 1132.

117. Lablab purpureus (L.) Sweet var. **lignosus** Prain (Fabaceae)

Local name : *Butta semi* (R)

Parts used : Fruits

All communities : Beans are used as vegetables. Bonda begin consuming this only after the festival of *Poos porv*

Voucher specimen	:	FMF 1205.

118. Lablab purpureus (L.) Sweet var. **purpureus** (Fabaceae)

Local name	:	*Mal semi* (R)
Parts used	:	Fruits
All communities	:	Beans are used as vegetables. Bonda begin consuming this only after the festival of *Poos porv*
Voucher specimen	:	FMF 428.

119. Lagenaria siceraria (Molina) Standl. (Cucurbitaceae)

Local name	:	*Thumpa, Thompa* (R)
Parts used	:	Fruits
All communities	:	Cultivated for its fruits that are used as vegetable; Gourds are used as vessels
Poraja	:	Gourds are used for making a single-stringed instrument called "*tola*"
Voucher specimen	:	FMF 1199.

120. Lagerstroemia parviflora Roxb. (Lythraceae)

Local name	:	*Amor* (B), *Cheyuni* (G)
Parts used	:	Wood and fruits
Bonda	:	Ripe fruits are ground, mixed with charcoal and applied on cattle wounds infected with worms (Mongla Muduli)
Gadaba	:	Wood is used for house construction and for making agricultural implements
Voucher specimen	:	FMF 789.

121. Lannea coromandelica (Houtt.) Merr. (Anacardiaceae)

Local name	:	*Kanbeli marnu, Palakara* (K), *Porontho* (B) *Pithmari* (R)
Parts used	:	Stem, bark and leaves

All communities	:	(i)	Stem is used as toothbrush
		(ii)	Wood is used to make yokes to carry weights and as firewood
Gadaba	:		Tender leaves are ground with that of *noondri* (*Syzygium cumini*) and *tendca* (*Woodfordia fruticosa*), and eaten for stomach ache (Sukra Dishari)
Poraja	:	(i)	Leaf paste is applied on boils and scabies
		(ii)	Root paste is consumed for stomach ache (Raghu Jani)
Bonda	:		Bark pieces are kept in the container used to collect toddy from *solap* (*Caryota urens*) to increase the potency of toddy
Voucher specimen	:		FMF 452.

122. Lantana camara L. (Verbenaceae)

Local name	:	*Poorda* (P, G), *Urli, Kikiri katta* (K)
Parts used	:	Whole plant and fruits
All communities	:	Plants are used as live fencing for home gardens
Kondh	:	Fruits are consumed for hardening stools (Kora Mali)
Voucher specimen	:	FMF 392.

123. Leucas aspera Link (Lamiaceae)

Local name	:		*Thummi* (P)
Parts used	:		Whole plant and leaves
Poraja	:	(i)	Leaf paste is applied for skin problems of cattle
		(ii)	A drop of leaf extract is inhaled through each nostril to cure migraine (Raghu Jani)
		(iii)	Whole plant is used in domestic rituals
Voucher specimen	:		FMF 466.

124. Limnophila conferta Benth. (Scrophulariaceae)

Local name	:		*Katho mali* (P), *Kadomali* (K), *Bangaru* (G)
Parts used	:		Whole plant and roots
Kondh	:		Whole plant is eaten with that of *Rotala illecibroides* to cure obesity
Gadaba	:	(i)	Plant is consumed as greens
		(ii)	Plant paste is applied on wounds in cattle caused by rubbing of the yoke
		(iii)	Roots are crushed and the paste is applied for toothache
Voucher specimen	:		FMF 727.

125. Litsea glutinosa (Lour.) C.B. Rob. (Lauraceae)

Local name	:	*Tangia achina* (R)
Parts used	:	Bark and gum
All communities	:	Gum oozing out from bark can stop bleeding from injuries and promote flesh formation
Poraja	:	Bark is ground with leaves of *billai gatta* (*Mimosa pudica*) and applied for injuries inflicted by weapons
Voucher specimen	:	FMF 1206.

126. Lycopersicon esculentum Mill. (Solanaceae)

Local name	:	*Tomat* (R)
Parts used	:	Fruits
All communities	:	Cultivated for its fruit that is consumed as vegetable. Seeds are purchased from the local market
Voucher specimen	:	FMF 1202.

127. Macrotyloma uniflorum (Lam.) Verdc. (Fabaceae)

Local name	:	*Kalot* (R)
Parts used	:	Seeds

All communities : (i) Cultivated as a pulse crop.

(ii) Gruel prepared by boiling the seeds in water, is supposed to be highly nutritive and is consumed by the new mothers of the tribal communities in the Koraput region. New mothers of the Kondh community are fed exclusively with this gruel for seven days after childbirth.

Voucher specimen : FMF 1189.

128. Madhuca longifolia (Koenig) Macbr. (Sapotaceae)

Local name : *Boa* (B)

Parts used : Fruits and seeds

Bonda : (i) Liquor is distilled from flowers and consumed by the Bonda community alone since the mature trees of this species are currently found only in the Bonda hills and its neighbourhood. These trees have been wiped out elsewhere

(ii) Edible oil is extracted from seeds

Voucher specimen : FMF 1191.

129. Mallotus philippensis (Lam.) Mull. Arg. (Euphorbiaceae)

Local name : *Sinthooro* (R), *Hindoori* (K)

Parts used : Wood and fruits

All communities : (i) Fruits are collected and sold as NTFP for dye

(ii) Wood is used as timber for house construction

Voucher specimen : FMF 197 (*fig.* 38).

130. Mangifera indica L. (Anacardiaceae)

Local name : *Maha* (K), *Maska* (P), *Ambo* (PI)

Parts used : Bark and seed kernels

All communities : Dried mango kernel from well mature mangoes are put in bamboo baskets and soaked in stream overnight to remove bitterness. They are ground

			in the next morning and made into porridge, alone or with flour of finger millet, and consumed. Bonda prefer to mix the soaked kernels with rice while cooking
Kondh	:		Bark is eaten with that of *jambo marnu* (*Syzygium cumini*) for dysentery
Gadaba	:		Mango kernels are ground and eaten with finger millet porridge for fever
Voucher specimen	:		FMF 1208.

131. **Maytenus hookeri** Loes. (Celastraceae)

Local name	:		*Dharini* (K), *Garud* (P)
Parts used	:		Bark
Poraja	:		Bark paste is applied on the patient to drive away evil spirits (Raghu Jani)
Voucher specimen	:		FMF 340.

132. **Melia azedarach** L. (Meliaceae)

Local name	:		*Garuda* (P)
Parts used	:		Roots and seeds
Poraja	:	(i)	The bitter seeds were used for snake bite (Raghu Jani)
		(ii)	Root paste is administered orally for stomach ache (Raghu Jani)
Voucher specimen	:		FMF 448.

133. **Merremia vitifolia** (Burm. f.) Hallier f. (Convolvulaceae)

Local name	:		*Behendi maal* (R), *Hingeri pakodi puyu* (K)
Parts used	:		Leaves
Kondh	:		Leaves are crushed and rubbed externally for stomach ache (Hori Hontal)
Voucher specimen	:		FMF 732 (*fig.* 39).

134. Michelia champaca L. (Magnoliaceae)

Local name	:		*Chompae* (R), *Chompa* (PI)
Parts used	:		Flowers and seeds
All communities	:		Flowers are used as adornment by girls
Bonda	:		Seed oil is applied on the body to repel mosquitoes and for fragrance
Voucher specimen	:		FMF 1187.

135. Mimosa pudica L. (Mimosaceae)

Local name	:		*Lemosac* (B), *Billai gatta* (P), *Gori saveli* (K), *Lach kutti* (R)
Parts used	:		Whole plant, roots, leaves and flowers
Bonda	:	(i)	Flowers are kept on the earlobe to ensure safe child delivery
		(ii)	Leaves are kept on the earlobe to induce sleep
Poraja	:	(i)	Leaves are ground with the bark of *tangia achina* (*Litsea glutinosa*) and applied for injuries inflicted by weapons
		(ii)	Roots are chewed with that of *surrikar* (*Tridax procumbens*) and also kept at the affected spot for gum ache (Raghu Jani)
Kondh	:		Roots are chewed for body swellings and also applied on the affected region
Gadaba	:	(i)	Plant is ground with jaggery and inflorescence of *sukhali butta* (*Bidens pilosa*) and given to the opposite sex, especially girls without their knowledge to make them amorous
		(ii)	Roots are eaten raw for stomach ache
Voucher specimen	:		FMF 477 (*fig*. 40).

136. Momordica charantia L. var. **charantia** (Cucurbitaceae)

Local name	:		*Karla* (R)
Parts used	:		Fruits

All communities : Cultivated for its fruits that are consumed as vegetable

Voucher specimen : FMF 1193.

137. Momordica charantia L. var. **muricata** (Willd.) Chakrav. (Cucurbitaceae)

Local Name : *Jaada Karla Maada* (K)

Parts used : Leaves and fruits

Kondh: (i) Leaves and fruits are used as vegetables

(ii) Leaf extract is mixed with mustard oil and applied on itches

Voucher specimen : FMF 630 (*fig*. 41).

138. Mucuna pruriens (L.) DC. (Fabaceae)

Local name : *Tuda* (K), *Kacheng* (G), *Rundul* (R)

Parts used : Roots and fruits

Kondh, Poraja and Gadaba : Stinging hairs of fruits are removed, the fruit powdered, and given to cattle for deworming

Gadaba : Roots are boiled with stream/river water and administered to males as tonic. It is believed to impart the strength of iron to the person who consumes it (probably an aphrodisiac)

Voucher specimen : FMF 222 (*fig*. 42).

139. Murraya koenigii (L.) Spreng. (Rutaceae)

Local name : *Mirchinga* (R), *Musinga* (B)

Parts used : Leaves

All communities : Leaves are used in curries

Voucher specimen : FMF 1186.

140. Murraya paniculata (L.) Jack (Rutaceae)

Local name : *Pondkada, Pondgal goch* (G), *Indramoris* (R)

Parts used : Bark

Gadaba : Bark paste is made into tablets and taken for

		three days for swellings on any part of the body along with jaggery (Sukra Dishari)
Voucher specimen	:	FMF 585.

141. Neolitsea foliosa (Nees) Gamble (Lauraceae)

Local name	:	*Pitti porodo* (K)
Parts used	:	Wood
Kondh	:	Wood is used for house construction
Voucher specimen	:	FMF 746.

142. Nicotiana tabacum L. (Solanaceae)

Local name	:	*Pogu* (R)
Parts used	:	Leaves
All communities	:	Leaves are used to make cigar. Cultivated as well as found as an escape.
Voucher specimen	:	FMF 1194.

143. Ocimum canum Sims (Lamiaceae)

Local name	:	*Patha tulsi* (P), *Lulising* (B)
Parts used	:	Inflorescence
Kondh	:	Inflorescence is kept inside homes, especially in shelves as an insect repellant
Poraja	:	Bunches of inflorescence are hung before cattle shed as an insect repellant
Bonda	:	Inflorescence is boiled in oil, cooled and the oil is applied to hair when visiting relative's family. This imparts fragrance to the hair and also wards off lice
Voucher specimen	:	FMF 624.

144. Oroxylum indicum (L.) Kurz (Bignoniaceae)

Local name	:	*Pappeni* (P), *Rachu peppeni* (K).
Parts used	:	Bark

Poraja : (i) Bark paste is applied for measles

(ii) Bark paste is given for snake bite; Bark is also bound just above the area of the bite to prevent circulation of venom (Bondu Doli)

(iii) Bark is used in magical rituals to drive away evil spirits

(iv) Bark paste is consumed to ease delivery

Kondh : (i) Bark paste is consumed for fits

(ii) Bark is eaten raw to reduce body pain

(iii) Bark is eaten as tonic

Voucher specimen : FMF 613 (*fig*. 43).

145. Oryza sativa L. (Poaceae)

Local name : *Dhano* (R)

Parts used : Grains

All communities : Grains from several cultivars of this plant form a staple food. Grains are also used to prepare a traditional beverage called *pendham*

Voucher specimen : FMF 1188.

146. Paederia foetida L. (Rubiaceae)

Local name : *Pith jada* (K), *Suso kooni* (B), *Goo mal* (P)

Parts used : Whole plant and leaves

Bonda : (i) Plant is used in domestic rituals along with *siripupa* (*Pogostemon benghalensis*)

(ii) Plant is used in rituals along with leaves and branches of *nbong thithing* (*Ardisia solonacea*)

Kondh : Whole plant is used in domestic rituals

Poraja : Fumes blown over a hot iron rod from the squashed leaves on to the patient to drive away evil spirits (Bondu Doli)

Voucher specimen : FMF 606.

147. Panicum miliaceum L. (Poaceae)

Local name	:	*Suan* (R)
Parts used	:	Grains
All communities	:	Cultivated for its grains that form staple food
Voucher specimen	:	FMF 1196.

148. Panicum miliare Lam. (Poaceae)

Local name	:	*Kangu* (R)
Parts used	:	Grains
All communities	:	Cultivated for its grains that form staple food
Voucher specimen	:	FMF 1190.

149. Pedilanthus tithymaloides Poit. (Euphorbiaceae)

Local name	:	*Amar, Omar* (P)
Parts used	:	Leaves
Poraja	:	Tender leaves are ground with that of *doodh mal* (*Cryptolepis buchananii*), fruits of papaya, *laden* (long black earthworms) and *jha* (earthworm) and eaten as galactagogue.
Voucher specimen	:	FMF 478.

150. Pennisetum hohenackeri Hochst. ex Steud. (Poaceae)

Local name	:	*Somp piri, Soom piri* (R)
Parts used	:	Whole plant
All communities	:	Grass is used for thatching the traditional houses
Voucher specimen	:	FMF 1200.

151. Peucedanum dhana C.B. Clarke var. **dalzellii** C.B. Clarke (Apiaceae)

Local name	:	*Patha dang* (P)
Parts used	:	Roots
Poraja	:	Roots are ground and applied externally on the navel region along with sesame oil for indigestion in infants (Bondu Doli)

Voucher specimen : FMF 622 (*fig.* 44).

152. Phoenix acaulis Buch.-Ham. ex Roxb. (Arecaceae)

Local name : *Sindhi* (R)

Parts used : Roots and fruits

All communities : Fruits are edible. Plants are burnt to roast the grub of the beetle *sindhi kida* (*Rhyncophorus ferrugenius*) that infect the stem and the grubs are consumed

Poraja : Roots are made into a paste with barks of *girli* (*Indigofera cassioides*), *jarri* (*Buchanania lanzan*), flour of *suan* (*Panicum miliaceum*), roasted and made into *uppuma* and eaten when new mothers feel pain in the navel region

Gadaba : (i) Roots are ground with that of *doodh mal* (*Cryptolepis buchananii*) and fruit of papaya and eaten as galactagogue

(ii) Roots are boiled and given or eaten with liquor as a tonic by new mothers

Voucher specimen : FMF 369.

153. Phoenix sylvestris (L.) Roxb. (Arecaceae)

Local name : *Ndoinda* (B), *Khojuri* (R)

Parts used : Leaves and fruits

All communities : Fruits are edible; less preferred

Bonda : Leaves were used to make wreath that were worn around the head of women. This practice has almost become extinct

Voucher specimen : FMF 1195.

154. Phyllanthus amarus Schum. & Thonn. (Euphorbiaceae)

Local name : *Thethlijad* (P), *Booi amla* (R)

Parts used : Shoot and leaves

Poraja : (i) Tender shoots and pepper are ground in

mustard oil and the paste is applied on the area affected by measles using feathers of hen (Raghu Jani)

(ii) Leaves are eaten to reduce swellings

(iii) Leaf extract is applied for pain and wounds (Bondu Doli)

Voucher specimen : FMF 473.

155. Phyllanthus emblica L. (Euphorbiaceae)

Local name : *Surrei* (P), *Amla* (R)

Parts used : Branches and fruits

All communities : Fruits are edible and also sold in the local market

Poraja and Gadaba : Pieces of branches are thrown behind by the lone traveller to prevent ghosts from following

Voucher specimen : FMF 1183.

156. Piper nigrum L. (Piperaceae)

Local name : *Golmari* (P), *Golomircha, Golmirch* (R)

Parts used : Fruits

Poraja : Pepper is ground with tender shoots of *booi amla* (*Phyllanthus amarus*) in mustard oil and applied on the area affected by measles with feather of hen

Kondh : Twenty-one seeds of pepper are chewed with three leaves of *pothru kaja* (*Bryophyllum calycinum*) like pan for watery diarrhoea

Bonda : Pepper fruits are ground with rhizome of *gimae sirasiv* (*Costus speciosus*) and eaten for cough (Mongla Muduli)

Voucher specimen : FMF 813.

157. Piper trioicum Roxb. (Piperaceae)

Local name : *Bon pipali* (R)

Parts used	:	Stems
Bonda	:	Stem pieces are tied to *salop* (*Caryota urens*) tree to prevent it from dying due to evil glance
Voucher specimen	:	FMF 804.

158. Plumbago zeylanica L. (Plumbaginaceae)

Local name	:	*Chitramool* (R)
Parts used	:	Roots and bark
Poraja, Gadaba and Kondh	:	Bark is crushed and added to *pendham* with that of Alstonia venenata to increase the potency of the brew
Poraja	:	Roots are tied as talisman to the right arm if the left testis is swollen in men and vice versa. Claimed to be quite effective
Voucher specimen	:	FMF 1182.

159. Plumeria rubra L. (Apocynaceae)

Local name	:		*Taka phul* (P), *Soitonga* (R)
Parts used	:		Bark and flowers
Gadaba	:		Bark extract is consumed as a mild laxative
Poraja	:	(i)	Trees are planted along the boundary of burial grounds in memory of one's mother. A practice not universally common to all Poraja villages.
		(ii)	Bark is mixed with that of *jilleni* (*Euphorbia nivulia*), ground, burnt, made into *pitta* (steam-cooked cake) with flour of finger millet and eaten for facial and abdominal swellings (Dolo Jimia)
Kondh	:	(i)	Flowers are used in the festival of *Mag porv* for decorations and in rituals. Considered as a symbol of purity
		(ii)	Bark paste is applied for skin ulcers in cattle
Voucher specimen	:		FMF 658 (*fig*. 45).

160. Pogostemon benghalensis (Burm.f.) Kuntze (Lamiaceae)

Local name	:		*Siripupa, Sirisipa* (B), *Balu butta* (R)
Parts used	:		Roots, leaves and flowers.
Bonda	:	(i)	Leaves are heated and rubbed on the body for body pain (Soma Sisa).
		(ii)	Flowers are used in domestic rituals with the plant of *suso kooni* (*Paederia foetida*)
		(iii)	Roots are ground with that of *kumbareng* (*Smilax perfoliata*) and the extract is consumed for diarrhoea combined with stomach ache
Gadaba	:		Leaves are used in rituals to ward off wild animals and evil spirits
Kondh	:		Leaves are crushed and the vapour from the extract is inhaled for cough
Voucher specimen	:		FMF 404.

161. Pogostemon stellatus (Lour.) Kuntze (Lamiaceae)

Local name	:		*Paaka puyu* (K), *Kirengeyser* (B), *Dongor tulsi* (R)
Parts used	:		Leaves and inflorescence
Kondh, Poraja and Gadaba	:		Inflorescence is smoked as cigar for fever
Kondh	:	(i)	Roots and inflorescence are ground and eaten for fits
		(ii)	Inflorescence is fumigated to drive off evil spirits
Bonda	:		Leaves are crushed and rubbed on body to repel mosquitoes
Gadaba	:		Inflorescence is fumigated in the chicken pen to drive away fleas
Voucher specimen	:		FMF 1197 (*fig*. 46).

162. Polyalthia cerasoides (Roxb.) Bedd. (Annonaceae)

Local name	:		*San saaj* (R)

Parts used	:	Stems
Gadaba	:	Branch piece is erected as stump, 15 cm above the ground in a corner as a ceremonial mark of house construction. This stump is also worshipped as family deity as long as the family resides in the house
Voucher specimen	:	FMF 1192 (*fig*. 47).

163. Polygonum chinense L. (Polygonaceae)

Local name	:	*Kodi golcha* (K)
Parts used	:	Roots and stems
Kondh	:	Roots and stem are ground and the extract is consumed for hand pain
Voucher specimen	:	FMF 1207 (*fig*. 48).

164. Pongamia pinnata (L.) Pierre (Fabaceae)

Local name	:	*Ramsey* (B), *Karanch* (R)
Parts used	:	Seeds
All communities	:	Oil obtained from the seeds is applied on body during winter to keep the body warm
Poraja	:	Oil is applied for scabies
Voucher specimen	:	FMF 449.

165. Psidium guajava L. (Myrtaceae)

Local name	:	*Jam goch* (R), *Lokha jam* (G)
Parts used	:	Fruits
All communities	:	Cultivated for its edible fruits. Also occurs as an escape. Fruits are often sold in the local markets
Voucher specimen	:	FMF 1181.

166. Pterocarpus marsupium Roxb. (Fabaceae)

Local name	:	*Bija* (R)

Parts used	:	Wood
All communities	:	Heartwood is used for making furniture, especially cots
Voucher specimen	:	FMF 1185 (*fig*. 49).

167. Pterospermum acerifolium (L.) Willd. (Sterculiaceae)

Local name	:	*Mochkund* (R)
Parts used	:	Flowers
Poraja and Kondh	:	Flowers are kept on the earlobes of women during delivery to ease pain
Kondh	:	Flowers are eaten raw with that of *tala* (*Alstonia scholaris*) for dysentery (Hori Hontal)
Gadaba	:	(i) Flowers are eaten raw for stomach ache
		(ii) Flowers are eaten raw to ease delivery
Voucher specimen	:	FMF 644 (*fig*. 50).

168. Pterospermum xylocarpum (Gaertn.) Santapau & Wagh (Sterculiaceae)

Local name	:	*Thimangalin* (B), *Mochkund* (R)
Parts used	:	Flowers
Kondh and Bonda	:	Flowers are kept on the ears of women during delivery to ease pain
Bonda	:	(i) Bark is kept in toddy from *solap* tree (*Caryota urens*) to increase the intoxicating potential of the brew
		(ii) Leaves are used in pollution ceremony after 21 days of childbirth
Voucher specimen	:	FMF 534 (*fig*. 51).

169. Rauvolfia serpentina Benth. ex Kurz (Apocynaceae)

Local name	:	*Poosi heeru* (K), *Gondalwa* (P), *Pathal gorud* (R)
Parts used	:	Roots and bark
Gadaba and Poraja	:	Roots are eaten raw for stomach ache

Poraja	:	(i) Bark is eaten for body pain
		(ii) Roots are ground and consumed once with a glass of water in empty stomach to cure stomach pain
Kondh	:	Roots are eaten raw for stomach ache
Voucher specimen	:	FMF 594 (*fig*. 52).

170. Remusatia vivipara (Roxb.) Schott (Araceae)

Local name	:	*Konja tahodi* (K), *Katta saru* (R)
Parts used	:	Yam
Kondh and Poraja	:	Yam paste is applied on septic wounds. Poisonous if consumed
Voucher specimen	:	FMF 720 (*figs*. 53 & 54).

171. Rhaphidophora pertusa (Roxb.) Schott (Araceae)

Local name	:	*Vali atta* (K), *Hadsang aeong* (B), *God sadni, Hath sadari* (R)
Parts used	:	Shoot and nodes
Kondh	:	Nodes are ground and applied for body pain
Poraja and Bonda	:	Stem paste is applied for fractures
Voucher specimen	:	FMF 667.

172. Ricinus communis L. (Euphorbiaceae)

Local name	:	*Kinthuru* (B), *Jada* (R)
Parts used	:	Seeds
Kondh and Bonda	:	Oil extracted from the seeds is used for body massage
Poraja and Gadaba	:	Oil extracted from the seeds is used for culinary purposes
Gadaba	:	Oil is used to light lamp in the *gunia*'s (lady ritualist) hand and swayed before the spirit-affected person. This is accompanied by stroking the person gently with flowers of *dragu* (*Argemone mexicana*)

Voucher specimen : FMF 258.

173. Rotala illecebroides (Arn.) Koehne (Lythraceae)

Local name : *Kothomali* (K), *Gundri* (G)

Parts used : Whole plant and leaves

Kondh : Whole plant is ground with that of *kado mali* (*Limnophila conferta*) and eaten to cure obesity

Gadaba : Leaf extract is poured into eyes for eye pain

Voucher specimen : FMF 725.

174. Rubia cordifolia L. (Rubiaceae)

Local name : *Chirumcher, Romaal, Qrra mal* (R) *Kanjaro veni* (K)

Parts used : Roots, bark and leaves.

Poraja : (i) Root powder is consumed for deworming in humans and in cattle

(ii) Bark is tied on waist, arm or neck as talisman for fever

(iii) Root paste is eaten for fits

(iv) Roots and leaves are ground and eaten for intermittent fever (Raghu Jani)

(v) Roots are eaten raw and the paste is applied on scrotum for pain (Raghu Jani)

(vi) Roots are eaten raw for chest pain (Dayanidhi Pujari)

Kondh : Leaves are thrown into the fields as a rat repellant

Voucher specimen : FMF 693.

175. Rubus ellipticus Sm. (Rosaceae)

Local name : *Kunupa* (K), *Nangseri bullu* (G), *Moisi koli* (R)

Parts used : Fruits

All communities	:	Fruits are eaten and also sold in the market
Voucher specimen	:	FMF 750.

176. Rubus niveus Wall. (Rosaceae)

Local name	:	*Dongor kunupa* (K), *Moisi koli* (R)
Parts used	:	Fruits
All communities	:	Fruits are eaten and also sold in the market
Voucher specimen	:	FMF 1180.

177. Sarcococca coriacea (Hook.) Sweet (Buxaceae))

Local name	:	*Odihitiri* (K)
Parts used	:	Bark
Kondh	:	Bark paste is applied on forehead for fever
Voucher specimen	:	FMF 781.

178. Schefflera stellata (Gaertn.) Harms (Araliaceae)

Local name	:	*Chuchjeli* (K), *Kooi marin* (G), *Galang* (B), *Nundool* (R)
Parts used	:	Stems
Gadaba	:	The Gadaba consider this species as an indicator of water and believe that the presence of this tree by the side of a stream will ensure water flow all through the year
Bonda	:	Stem is used as sceptre while dancing and during the festival of *Choith porv*. Male members of the tribe can be seen carrying this sceptre while wandering from village to village for romance
Voucher specimen	:	FMF 654.

179. Schleichera oleosa (Lour.) Oken (Sapindaceae)

Local name	:	*Pooji* (P), *Kosomi marnu* (K), *Imbang* (B), *Koshum* (G)
Parts used	:	Wood, bark, fruits and seeds

Poraja, Gadaba and Kondh	:	Oil is extracted from seeds
Poraja and Gadaba	: (i)	Wood is used for making plough bases, paddy threshing pestles and yokes to carry weight
	(ii)	Fruits are edible
	(iii)	Oil is applied for body pain and sprains
Kondh	: (i)	Bark extract is applied for sprain
	(ii)	Oil is applied for scabies and wounds
Voucher specimen	:	FMF 462.

180. Selaginella nairii Dixit (Selaginellaceae) (Pteridophyte)

Local name	:	*Sapta* (B)
Parts used	:	Whole plant
Bonda	:	Roots are squashed and consumed for snake bite (Suma Bodnaik, Suma Dora)
Voucher specimen	:	FMF 709.

181. Semecarpus anacardium L.f. (Anacardiaceae)

Local name	:	*Bonwalia, Balia* (R)
Parts used	:	Pseudocarps and nut
Poraja and Gadaba	: (i)	Pseudocarps are edible
	(ii)	Oil from the pericarp of the nut is applied for minor bruises
	(iii)	Oil from the pericarp of the nut is applied on foot cracks
Gadaba	:	Pseudocarps are stalked in sticks, dried, fried and eaten
Voucher specimen	:	FMF 1184 (*fig*. 55).

182. Senecio nudicaulis Buch.-Ham. ex D. Don (Asteraceae)

Local name	:	*Palu kuna* (K)
Parts used	:	Roots

Kondh	:	Roots are eaten raw as well as smashed and applied externally on the breasts as galactagogue
Voucher specimen	:	FMF 1179.

183. **Senna hirsuta** (L.) H.S. Irwin & Barneby (Caesalpiniaceae)

Local Name	:	*Chakunda* (R)
Parts used	:	Roots
Kondh	:	Root paste is applied on wounds that persist for years (Hori Hontal)
Voucher specimen	:	FMF 631.

184. **Sida acuta** Burm. f. (Malvaceae)

Local name	:	*Jaati batni* (P), *Kadbil* (G), *Garada* (K)
Parts used	:	Whole plant
All communities	:	Plant is used as broom to clean thrashing grounds and for areas other than houses
Voucher specimen	:	FMF 245 (*fig.* 56).

185. **Sida cordata** Burm. f. (Malvaceae)

Local name	:	*Eda adi cher* (G), *Sirsi* (B)
Parts used	:	Leaves
Gadaba	:	Leaves are ground with that of *limba* (*Azadirachta indica*) and the paste is applied for skin diseases, mostly in private parts (Sukra Dishari)
Bonda	:	Leaves and roots are ground and applied externally for all sorts of pain
Voucher specimen	:	FMF 363.

186. **Smilax perfoliata** Lour. (Smilacaceae)

Local name	:	*Paraska mada* (K), *Muthirimal* (R) *Kumbareng* (B)
Parts used	:	Roots, stems and leaves

Kondh and Gadaba	:	Stems are tied to the trunk of *solap* tree (*Caryota urens*) to increase the flow of sap
Gadaba and Poraja	:	Twigs are used as toothbrush
Kondh	:	Leaves and bark are given for mental illness
Gadaba	:	Stem pieces are tied above the door to ward off evil spirits and to promote peace at home. The plant is supposed to be collected for this purpose only on an auspicious star called *onradha*
Bonda	:	(i) Roots are ground with that of *siripupa* (*Pogostemon benghalensis*) and the extract is consumed for diarrhoea combined with stomach ache
		(ii) Stems with mud from the wayside are used in magic rituals to ward off spirits
Poraja	:	Stems are buried in a pit inside the home, just near the door, to prevent entry of spirits. Sometimes also hung on roof
Voucher specimen	:	FMF 614.

187. Solanum melongena L. (Solanaceae)

Local name	:	*Baigun* (R)
Parts used	:	Fruits
All communities	:	Cultivated for its fruits. Fruits are used as vegetable either fresh or dried
Voucher specimen	:	FMF 1178.

188. Solanum torvum Sw. (Solanaceae)

Local name	:	*Porpaja, Gadhri phul* (P), *Poyya, Kodru poyya* (K)
Parts used	:	Roots, leaves, flowers and fruits
Poraja	:	(i) Leaves and fruits are eaten raw for fits
		(ii) Root is tied as talisman on women and children with an irritating character

Kondh	:	(i) Root paste is given for mental illness
		(ii) Flowers are ground and rubbed on stomach for indigestion and flatulence (Hori Hontal)
Voucher specimen	:	FMF 199 (*fig*. 57).

189. Sorghum bicolor (L.) Moench (Poaceae)

Local name	:	*Kaet jona* (R)
Parts used	:	Grains
All communities	:	Grains form one of the traditional staple foods. Its cultivation has shrunk in the recent past due to changing food preferences
Voucher specimen	:	FMF 2037.

190. Stachytarpheta urticifolia Sims (Verbenaceae)

Local name	:	*Ondra bathra* (P), *Porrenga butta* (K)
Parts used	:	Leaves
Poraja	:	Leaf paste is applied on wounds, especially those occurring beneath the finger and toe nails. Its related species *bathra* (*Stachytarpheta jamaicensis*) is considered as its female counterpart and is considered as "useless"
Kondh	:	Leaf extract is used as styptic to arrest bleeding from injuries
Voucher specimen	:	FMF 1177.

191. Stereospermum colais (Buch.-Ham. ex Dillwyn) Mabb. (Bignoniaceae)

Local name	:	*Baerthey maaling* (G)
Parts used	:	Fruits
Gadaba	:	Fruits are considered as a potential drug against snake bite. However no knowledge holder interviewed had used it for the purpose before
Voucher specimen	:	FMF 1176.

192. Syzygium cumini (L.) Skeels (Myrtaceae)

Local name : *Noondri* (P), *Jambo marnu* (K), *Jam Kuli* (G)

Parts used : Wood, bark, leaves and fruits

Kondh, Poraja and Gadaba : (i) Fruits are eaten as well as sold in the local markets

(ii) Wood is used for making plough handles

(iii) Wood is used for house construction

Gadaba : Tender leaves with that of *pithmari* (*Lannea coromandelica*) and *tendca* (*Woodfordia fruticosa*), are ground and eaten for stomach ache (Sukra Dishari)

Kondh : Bark is chewed with that of *ambo* (*Mangifera indica*) to stop dysentery

Poraja : (i) Bark paste is eaten for dysentery

(ii) Bark with that of *kumi* (Gardenia latifolia), *jarri* (*Buchanania lanzan*), *saaj* (*Terminalia tomentosa*) and coconut gratings are ground using stone and the extract is consumed for dysentery

(iii) Bark extract and tender leaf extract of *Woodfordia fruticosa* are consumed for stomach ache (Jogu Jani)

Voucher specimen : FMF 457.

193. Tamarindus indica L. (Caesalpiniaceae)

Local name : *Thimthim* (B), *Thenthuli* (R)

Parts used : Fruits

All communities : Pulp of the fruit is edible and also used in curries

Bonda : (i) Liquor is distilled from the fruits

(ii) Three-year-old stored fruits are mixed with plant extract of *nisudang* (*Musa balbisiana/M. paradisiaca*) and administered as emetic for drunken people (Mongla Muduli)

Voucher specimen : FMF 521.

194. Terminalia bellirica (Gaertn.) Roxb. (Combretaceae)

Local name : *Tandri marnu* (K), *Billa* (R)

Parts used : Fruits

All communities : Fruits are used as a component of the Ayurvedic *triphala* and also sold as a NTFP for the same purpose

Voucher specimen : FMF 302.

195. Terminalia chebula Retz. (Combretaceae)

Local name : *Horda* (R)

Parts used : Fruits

All communities : Fruits are used as a component of the Ayurvedic *triphala* and also sold as NTFP for the same purpose.

Poraja : (i) Fruit is powdered with leaves of *ganjai* (*Cannabis sativa*) and consumed for cough (Dayanidhi Pujari)

(ii) Fruits are burned and the ash is applied externally for fractures (Raghu Jani)

Kondh : The epicarp and mesocarp of the fruit are burnt, powdered and consumed along with water for cough and cold

Voucher specimen : FMF 467.

196. Terminalia tomentosa Wight & Arn. (Combretaceae)

Local name : *Marding* (G), *Saaj* (R)

Parts used : Stem, bark and leaves

Gadaba : (i) Wood is used for house construction

(ii) Stem is used as toothbrush to cure headache

Poraja : (i) Bark paste is applied for bone fracture (Raghu Jani)

(ii) Bark with that of *kumi* (*Gardenia latifolia*), *jarri* (*Buchanania lanzan*), *noondri* (*Syzygium cumini*) and coconut gratings are ground using stone and the extract is consumed for dysentery (Raghu Jani)

(iii) Bark and leaves are added to liquor to increase the intoxicating potency

Voucher specimen : FMF 657.

197. Thalictrum foliolosum DC. (Ranunculaceae)

Local name : *Pitha sago* (R)

Parts used : Whole plant

Bonda : Plants are cooked and eaten as greens for stomach ache (Mongla Muduli)

Voucher specimen : FMF 800.

198. Themeda triandra Forssk. (Poaceae)

Local name : *Palda* (R)

Parts used : Whole plant

All communities : Plants are used for thatching the traditional houses

Voucher specimen : FMF 1175.

199. Thunbergia fragrans Roxb. (Acanthaceae)

Local Name : *Doodh mal* (P), *Pakodi bandh* (K), *Romal phul* (R)

Parts used : Roots and flowers

All communities : Flowers are tied to the inflorescence stalk of *solap* tree (*Caryota urens*) to increase the yield of toddy

Gadaba : Root paste is eaten for stomach problems

Poraja : Root powder is mixed with flour of finger millet, made into a cake and given to women to stop blood discharge in urine (Raghu Jani)

Voucher specimen : FMF 465 (*fig*. 58).

200. Thysanolaena maxima (Roxb.) Kuntze (Poaceae)

Local name : *Heppori* (K), *Phul batni* (R)

Parts used : Roots and inflorescence

All communities : Bundle of inflorescences is used as broom to clean houses

Kondh : (i) The mature inflorescence stalks are shade dried and reinforced with mud to provide an insulation layer beneath the asbestos and tiled roofs

(ii) Roots and leaves are ground and eaten for fits (Dona Jani)

(iii) Root paste is applied for body swelling

Voucher specimen : FMF 378.

201. Tragia involucrata L. (Euphorbiaceae)

Local name : *San lachuani* (R)

Parts used : Seeds and leaves

Poraja : Leaves are given for cattle deworming (Raghu Jani)

Kondh : Seeds are smashed and rubbed on forehead for headache

Voucher specimen : FMF 655.

202. Tridax procumbens L. (Asteraceae)

Local name : *Mitti morado* (K), *Ventheri* (G), *Surrikar* (P)

Parts used : Roots and leaves

Kondh and Poraja : Leaf extract is used as antiseptic for wounds

Gadaba : Leaf extract is applied on bruises on legs of cattle

Poraja : Root is chewed with that of *Billai gatta* (*Mimosa pudica*) for gum ache (Bondu Doli)

Voucher specimen	:	FMF 598.

203. Urena lobata L. (Malvaceae)

Local name	:	*Sicajal* (P)
Parts used	:	Whole plant
Poraja	:	Plants are used in magical rituals, to cure stomach ache due to evil spirits
Voucher specimen	:	FMF 620.

204. Vicoa auriculata Cass. (Asteraceae)

Local name	:	*Thusibulaing* (B)
Parts used	:	Roots
Bonda	:	Root paste is applied on forehead for headache (Suma Bodnaik, Suma Dora)
Voucher specimen	:	FMF 718 (*fig.* 59).

205. Vigna mungo (L.) Hepper (Fabaceae)

Local name	:	*Biri* (R)
Parts used	:	Seeds
All communities	:	Cultivated for its seeds. The seeds are stored after mixing with wood ash. *Biri* forms an ingredient of a variety of traditional recipes
Voucher specimen	:	FMF 1174.

206. Vigna radiata (L.) R. Wilczek (Fabaceae)

Local name	:	*Dongarani* (R)
Parts used	:	Seeds
All communities	:	Cultivated for its seeds. *Dongarani* forms an ingredient of a variety of traditional recipes
Voucher specimen	:	FMF 1173.

207. Wendlandia glabrata DC. (Rubiaceae)

Local name	:	*Raakeyee* (K)

Parts used	:		Wood
Kondh, Poraja and Gadaba	:		Wood is used for house construction and for making furniture. Wood is resistant to termite and other infection and is valued high
Voucher specimen	:		FMF 735.

208. Woodfordia fruticosa (L.) Kurz (Lythraceae)

Local Name	:		*Dokrioda* (P), *Gisangthae*b (B), *Tendca* (R)
Parts used	:		Leaves and flowers
Poraja	:	(i)	Flowers are roasted and eaten as food. Flowers are also made into curry
		(ii)	Leaf extract is applied for wounds near the ear
		(iii)	Leaves are fried in oil from *olsi* (*Guizotia abyssinica*), made into a paste and applied on wounds near the ear
		(iv)	Tender leaves are eaten with bark of *noondri* (*Syzygium cumini*) for stomach ache (Jogu Jani)
Gadaba	:		Tender leaves with that of *pithmari* (*Lannea coromandelica*), *jam gocho* (*Syzygium cumini*) are ground and eaten for stomach ache (Sukra Dishari)
Voucher specimen	:		FMF 460 (*fig.* 60).

209. Xylia xylocarpa (Roxb.) W. Theob. (Mimosaceae)

Local name	:	*Tangini* (K)
Parts used	:	Seeds
Kondh	:	The seeds are roasted over fire and consumed. Said to be mildly sweet. According to sources, the tree was once common but presently represented by a single tree in Kondh Pungar
Voucher specimen	:	FMF 1172 (*fig.* 61).

210. Zanthoxylum armatum DC. (Rutaceae)

Local name	:	*Rongwali* (P), *Kathiri* (K), *Rongbeli* (R)

Parts used	:	Bark
Poraja, Kondh and Gadaba	:	Bark is powdered and used as tooth powder for toothache
Kondh and Gadaba	:	Bark powder is applied for infection of hoof in cattle
Kondh	: (i)	Bark powder is sprinkled on stagnant water to stupefy fish
	(ii)	Bark is soaked in water for three days and sprayed on vegetables as bio-insecticide (This knowledge could have been introduced by the NGOs and governmental agencies concerned with bio-farming, as reported by the villagers)
Voucher specimen	:	FMF 368.

211. Ziziphus jujuba Mill. (Rhamnaceae)

Local name	:	*Gameda* (R)
Parts used	:	Leaves
Poraja	:	Leaves when squashed form lather which is applied externally on face and the abdominal region to reduce swellings
Voucher specimen	:	FMF 1171.

212. Ziziphus rugosa Lam. (Rhamnaceae)

Local name	:	*Onungo marnu* (K), *Kontakolli, Kotkolli* (R), *Surrei bullu* (G)
Parts used	:	Stem, bark, leaves, fruits and seeds
All communities	: (i)	Fruits are edible
	(ii)	Wood is used as firewood
	(ii)	Plant is used for live fencing
Kondh	: (i)	Bark extract is applied as antiseptic for cuts and wounds
	(ii)	Bark paste is consumed for snake bite (Doni Jani)

	(iii)	Bark extract is consumed for common cold (Doni Jani)
Gadaba	:	Three leaves are pinned to broom and a person lacking appetite is made to spit on it thrice and the broom is left near the jungle. This ritual is believed to restore normal appetite
Poraja	:	Seeds are dried and used as substitute for *kandul* (*Cajanus cajan*) in curries
Voucher specimen	:	FMF 751.

4

Ethnotaxonomy of Plants

Classification . . . is an absolute and minimal requirement of being or staying alive
— Simpson 1961

CLASSIFICATION forms the basis of knowledge. It simplifies the process of understanding complex phenomena or processes by categorising them. Besides, folk classification helps in easy retrieval of information on local resources. It is also essential for the utilisation and manipulation of natural resources (Berlin 1992; Souza and Begossi 2007). A majority of plants included under folk classification system is generally believed to be plants of high utility value (Voeks and Nyawa 2006). However, plants that are normally considered as of less utility value by the researcher may be accorded high value, mostly cultural by the community and hence may figure prominently in the classification system (Hunn 1982). Anonby (2006) has shown that folk classifications accept the usage of many synonyms to denote a single species which is not tolerated by the Western system of classification. Newmaster et al. (2007) have shown that folk classification can be more robust than modern classification systems. This shows that folk classification is quite important from the community's point of view.

Berlin et al. (1974) have developed a classification model based on Tzeltal folk classification which largely holds good for most of the biological folk taxonomical systems. However, it might not be appropriate for non-biological classifications (Holman 2005). The chief aspects of their classification model are explained below.

Berlin et al. have found out that the folk classification systems are more or less similar in most of the indigenous communities. The system is found to be comprised of several taxonomic hierarchies such as the unique beginner, life form, genera, species and variety.

Berlin's Folk Classification

UNIQUE BEGINNER

This category is an all-inclusive category and members of this category are seldom named. More widely used unique beginner categories are "plants" and "animals". However, most of the tribal communities do not refer to these inclusive categories by a single linguistic expression, equivalent to the English terms.

LIFE FORMS

This category mostly comprises taxa that are descriptive of the plant habit in nature, e.g. trees and climbers. Linguistically, the taxa of this category are labelled with simple lexemes.

GENERIC

Taxa of this category are mostly labelled with simple lexemes. The lexemes may be of both productive as well as unproductive type. The generic taxa may or may not be included under the life form category.

SPECIES

The taxa of species category are less in number when compared with those of generic category. They are essentially of secondary lexeme in nature. One of the two epithets of the name is usually of the "genus", e.g. wood apple is a folk species of apple.

VARIETY

They are sub-specific in nature and can be usually recognised by a name of three epithets.

Besides these, Berlin also recognises another category called "Intermediate taxa or covert categories" that are directly included under the category "life form".

Generally, folk classifications fall into the natural systems of classification with their system of nomenclature rooted mostly in characters such as utility, morphology, ecology and chemistry (Kakudidi 2004). In the Koraput region too, the system of plant nomenclature is based on characters such as utility value, morphology, ecology, quality, and beliefs attached. The study of ethnotaxonomy, along with the folk system of nomenclature would enable better understanding of the community's knowledge system and the patterns. The ethno-taxonomy and nomenclature of the Kondh, Poraja, Gadaba and Bonda are discussed below.

Kondh Taxonomical System

UNIQUE BEGINNER

The category "unique beginner" explained by Berlin is not found. In other words, the Kondh do not have a term equivalent to the English term of "animal" or "plants". The word commonly used is *jotho* which is the local adaptation of the Odia word *janthu* and hence is not essentially Kuvi.

LIFE FORM

Five types of life forms are recognised by the Kondh, viz. *marnu* (trees), *gocha matta* (bushes), *mada* (creeper/climber), *podhi* (plants growing up to 1.5 ft) and *puyu* (plants with conspicuous flowers). The grouping is on the basis of height of the plant except for the category *puyu*. These life form categories can be classified broadly into two types, the recognised and non-recognised. *Marnu*, *mada* and *puyu* fall into the recognised type where the members affiliated to this category are tagged with the epithet — *marnu*, *mada* or *puyu*. The other two categories — *gocha matta* and *podhi* — are not found to have members with the category name tagged to the lexeme. However, they are used to group plants. The life form *puyu*, which comprises plants with conspicuous flowers is unique to the Kuvi system of classification. The conspicuousness used as criteria to recognise such taxa are mostly bright colours or unique shapes. Example: *Paaka puyu* (*Pogostemon stellatus*).

GENERA

The taxa belonging to the generic category are larger in number than the life form category. Most of the generic names are found tagged with the epithet corresponding to the life form under which it is included. Examples of few genera are *paalu mada* (*Cryptolepis buchananii*), *jam* (*Syzygium cumini*), *heppori* (*Sida acuta*), *kutcha* (all edible greens), *oso mada* (*Argyreia nervosa*), *paraska mada* (*Smilax perfoliata*), *karla mada* (*Momordica charantia*), *kunuva* (*Murraya koenigii*).

SPECIES

The Kuvi species are included under the generic category and are either tagged with the epithets corresponding to their respective genera and also their life form category or tagged with their corresponding generic category alone. Examples of some Kuvi species are, *san loho mada* (*Galactia tenuiflora*), *podho oso mada* (*Pueraria tuberosa*), *kotti kutcha* (*Edible ferns*), *attuva kutcha* (*Boerhavia diffusa*), *pennu heppori* (*Asparagus racemosus*), *odi jam* (*Chionanthus mala-elengi*), *monji jambo marnu* (*Psidium guajava*), *jada karla mada* (*Momordica charantia* var. *muricata*), *kaja kunuva* (*Clausena excavata*).

Nomenclature and Recognition of Species by the Kondh

The species category in Kuvi system of classification is done in comparison with the folk types. Folk types are genera that extend to represent an individual species too. For instance, *jam* is a genus under which *odi jam* and *monji jambo marnu* are grouped. However, the genus *jam* itself extends to represent an individual species too. Folk types act as templates for grouping allied species. Such grouping may be based on various mechanisms such as morphology, utility value, ecology, beliefs, etc. It should be noted that the name *jam* is commonly used by all communities in the Koraput region. Yet it is treated as Kuvi, since the species attributed to the genera are tagged with Kuvi epithets.

Sometimes, two characters may be used parallel or linearly to group species. *Jam* (*Syzygium cumini*) is the folk type for grouping *odi jam* (Bear *jam*, *Chionanthus mala-elengi*) and *monji jambo marnu* (Many-seeded *jam*, *Psidium guajava*). Here grouping is done on the basis of two characters, viz. morphology and utility. *Jam* is related to *manji jambo marnu*, but differs mainly in the many-seeded character. However, the same character cannot be used to differentiate *jam* and *odi jam* due to the single-seeded nature in both these plants. Hence, a second character-utility value is employed here. *Jam* is edible while *odi jam* is not edible and can be eaten only by bears. Thus we find two characters used parallely at the same time to species of the same genus.

Similarly, *loho mada* (iron climber) and *san loho mada* belong to the same genus of *loho mada*. Here, *loho mada* is the folk type. Here too, grouping seems to be done on the basis of two characters, but not distinctly evident. *Loho mada* refers to iron creeper/climber (*Hiptage benghalensis*) while *san loho mada* refers to smaller iron creeper (*Galactia tenuifolia*). Looking from a utility point of view, the name reflects the stronger nature of cordage obtained from *H. benghalensis* and the cordage of lesser strength obtained from *G. tenuifolia*. The name also reflects the smaller stature of *G. tenuifolia*.

ECOLOGY

Ecology plays the important role in grouping *karla mada* (bitter creeper, *Momordica charantia*) and *jada karla mada* (jungle bitter creeper, *Momordica charantia* var. *muricata*) together. Here the fact that *jada karla mada* is found in the jungle alone and not cultuivated is used to distinguish it as a separate species from a similar-appearing *karla mada*.

MORPHOLOGY

Kunuva (*Murraya koenigii*) and *kaja kunuva* (*Murraya paniculata*) are distinguished as separate species on the basis of leaf size. The larger leaves of *M. paniculata* are used to distinguish it from *M. koenigii*. here, *M. koenigii* is the folk type.

Bonda Taxonomical System

UNIQUE BEGINNER

Just like the Kuvi system of nomenclature, the Bonda system too lacks an equivalent term for unique beginner.

LIFE FORM

The Bonda category of life form is not visible, since the epithet corresponding to the life form of which the genera are inclusive is not tagged to the genera. The Bonda also recognise mushrooms and free-floating algae as separate life forms. The life forms recognised are, *simo* (trees), *latta* (bushes), *dalai* (herbs growing up to 1.5 feet), *dailey* (climbers), *kirey* (grass), *gothi sabur* (plants using other plants for support), *nthooi* (mushrooms), *sailom* (free-floating algae).

GENERA

The Bonda classification recognises taxa up to the species level but doesn't recognise any genera, except for a unique genus *sing* under which species such as *jali sing* are recognised. Sing denotes "sun" and there appears to be no common linkage between species of these genera other than their height, i.e. they are plants growing up to 3 ft only. It should also be noted that *sing* which corresponds to the genus occurs as a suffix and not prefix unlike other communities. However, the apparent existence of species such as *bon pipali,* belonging to *pipali* are not considered for study due to their Odia origin.

SPECIES

As discussed above, the taxa recognised as species are included directly under the category of life form except one group that is organised under the genus "Sing". Examples of species of *sing* are, *jalising* (ferns), *lulising* (*Ocimum canum*), *kooisesing* (*Crotalaria pallida*), *baasing* (*Desmodium triquetrum*). Examples for some species included directly under the life form category in the Bonda system are: *siri pupa* (*Pogostemmon benghalensis*), *goising koi* (*Canscora diffusa*), *nbong thithing* (*Ardisia solonacea*), *suso kooni* (*Paederia foetida*), *pudubullu* (*Bryophyllum calycinum*), etc.

Nomenclature and Recognition of Species by the Bonda

Species are mostly recognised individually. This is evident from the absence of any genera other than *sing*. But the nomenclature system is clearly based on mechanisms such as utility value, ecology, morphology and quality.

UTILITY

Some of the plants named on the basis of utility value are, *emur thuthur* (plant to treat pain in a small area, *Curculigo orchioides*), *gising koi* (*Canscora diffusa*, plant to treat diseases in fowls), and *kirengeyser* (mosquito repellant, *Pogostemon stellatus*).

ECOLOGY

Siripupa (*Pogostemon benghalensis*) is an example for plants named on the basis of its ecology. *Siripupa* can be translated as a "plant found in low lying areas", evidently reflecting the ecology.

MORPHOLOGY

Examples for plants named on the basis of their morphological characters are, *baasing* (*baa*=sides; refers to the winged petiole, *Desmodium triquetrum*), *thimangalin* (tall tree, *Pterospermum xylocarpum*), *nthoba* (plant with fruits resembling fish eggs, *Rubus spp.*)

QUALITY

The quality or nature of the plant is also used in naming them, e.g. *pudubullu* (*Bryophyllum calycinum*, makes put-put sound), *gana* (plant that itches, *Ficus asperrima*), *lemosac* (sleeping plant, *Mimosa pudica*)

Gadaba Taxonomical System

The classification system could not be studied completely due to the language erosion. All villages studied, except Dokriguda, were using Odia names mostly to refer to their plants. Such plant names were not considered. Example: *jam* (*Syzygium cumini*); *lokha jam* (*Psidium guajava*), *saaj* (*Terminalia tomentosa*); *san saaj* (*Polyalthia cerasoides*), etc.

UNIQUE BEGINNER

The Gadaba system of classification does not have a term equivalent to the English word "plants". However, unlike the other three communities, they do recognise the unique beginner category animals by the term *jeebeyl* which is a pure Gadaba word.

LIFE FORMS

The Gadaba recognise four life form categories, viz. *marin* (tree), *butta* (bush), *theeg* (climber) and *bullu* (plants with edible fruits). The life form category *bullu* which stands for plants with conspicuous and generally edible fruits is unique to the Gadaba system of classification. Plants recognised under this category are usually known to be tagged with the term *bullu*. *Bullu* seem to run parallel to the *puyu* of Kondh, e.g. *surrei bullu* (*Zizyphus rugosa*), *nangseri bullu* (*Rubus spp.*)

GENERA

Examples of some generic categories are *koori butta* (*Asparagus racemosus*), *thad marin* (*Caryota urens*), *koodoi* (*Holarrhena antidysenterica*), *sukhali butta* (*Bidens pilosa*), *kokkoon* (mushrooms).

SPECIES

The category of species could not be well studied due to language erosion. However, there seems to have existed a well-defined system of classification that extended till the species level. Example of a species still recognised is *bod koodoi* (*Alstonia venenata*), *kant kokoon* (leafy mushrooms, crustose lichens), *bomso kokoon* (mushrooms growing on bamboos).

Nomenclature and Recognition of Species by the Gadaba

Though the Gadaba seem to have a well-defined classification and nomenclatural system, we could only trace a few patterns of nomenclature due to language loss. Names originating on the basis of morphological understanding and quality were collected. The Gadaba system also uses folk types to recognise species. For instance, *koodoi* (*Holarrhena antidysenterica*) is used as folk type to recognise *bod koodoi* (*Alstonia venenata*). Some of the mechanisms used to name plants in the Gadaba system are explained below.

MORPHOLOGY

The name *suithi kanda* (*Cissampelos pareira*) qualifies the climber's stem which is compared to the line formed by the movement of ants.

Ficus semicordata which has the unique character of bearing "fruits" on the prop roots is referred as *vaerthey* (plant that has got unique roots).

The name *kant kokoon* (leafy mushroom) refers to all crustose lichens which are distinguished as different from the other mushrooms in being leafy or leaf-like.

QUALITY

Sukhali butta (plant that sticks, *Bidens pilosa*) is named after the seeds of the plant that stick to human beings for dispersal. This property also qualifies the plant to be used as a love potion.

Poraja Taxonomical System

A good majority of the plant names recorded from the Poraja villages are local Odia names (*adu jam* and *patha dang* are exceptions). This was the outcome of language erosion and it made it impossible to study the Poraja system of classification. However, a superficial idea obtained is given below.

UNIQUE BEGINNER

No unique beginner category is recognised in the Poraja system of classification.

LIFE FORMS

The three life forms recognised are *gocho* (trees), *butta* (bush) *mal* (climber), *jaati* (herbs). However, the terms are of local Odia in nature and the system too is common to the local people. Table 12 gives the terms used to denote the various life forms by all the communities.

GENERA

A few examples for the genera recognised by the Poraja are, *girli* (*Indigofera cassioides*), *kendu* (*Diospyros melanoxylon*), *batni* (*Sida acuta*), *limb* (*Azadirachta indica*), *amla* (*Phyllanthus emblica*).

SPECIES

Examples for taxa included in the species category are, *hathi girli* (*Breynia retusa*), *makorkendu* (*Diospyros sylvatica*), *phul batni* (*Thysanolaena maxima*), *dev baatni* (*Asparagus racemosus*), *booi limb* (*Andrographis paniculata*), *booi amla* (*Phyllanthus amarus*).

Table 12: Life Forms Category of the Four Communities

Community	**Trees**	**Shrubs/ Bushes**	**Climber/ creeper**	**Herbs**	**Grass**	**Others**
Kondh	*Marnu*	*Gocha matta*	*Mada*	*Podhi*		*Puyu* (flowers)
Bonda	*Simo*	*Latta*	*Dailey*	*Dalai*	*Kirey*	*Gothi sabur* (plants using other plants for support), *nthooi* (mushrooms), *sailom* (free-floating algae)
Gadaba	*Marin*	*Butta**	*Theeg*			*Bullu* (fruits)
Poraja	*Gocho**	*Butta**	*Mal**			

Note: * Denotes Odia names

Local Names, Their Origin and Implications for Loss of Language

Of the total number of names used to identify plants, 125 names are of regional and nine names are pan-Indian in nature. Poraja and Gadaba are the communities mostly using regional names instead of names from their own languages. This shows that the communities have lost the names in their own respective languages and have borrowed the regional names from Odia and pan-Indian names from Hindi. This has got serious implications since it is a well-agreed fact that language erosion leads to erosion of ecological knowledge and culture (UNESCO 2003). From Bernard 1992, we could clearly understand that the loss of their language is seen as a consequence of the loss of their original ethnic and cultural identity and cannot be stemmed unless the cultural identity is revived. The fact that language is the foremost intangible heritage on which other intangible heritages such as culture, knowledge and ethnicity have been built upon has to be highlighted. This is more relevant for these communities since Parji, Gadaba, Kuvi and Remo spoken by the Poraja, Gadaba, Kondh and Bonda communities respectively have been identified as "endangered languages" by the UNESCO in 2010.

5

Plants and Concepts in Healing

Ooyo aje leaf chase away for us the disease out of the flank.,
Awusa leaf heal the flank disease so that it may go away,
Agbe feather carry that flank disease away,
Aluko feather pick away the disease out of the flank.

— Sofowara 1982

PLANTS were the first found source of medicines in the history of human beings. Throughout the world, all societies are known to have relied upon plants at least during the initial stages of their formations. With the development of synthetic chemical-based allopathic system of medicine, the developed nations moved towards abandoning their plant-based system of medicine. Yet, they are also dependent on plants for deriving the active components that form a part of allopathic system of medicine. Currently, at least 25 per cent of the drugs used in the allopathic system of medicine are derived from plants (Kala & Sajwan 2007).

In spite of the massive popularity of allopathic system of medicine, indigenous communities all over the world continue to rely on their own predominantly plant-based systems of medicine. Perhaps, the indigenous societies found the allopathic system of medicine too impersonal to replace their indigenous systems of medicine (Vandebroek et al. 2004). It was the colonial rule, under which most of today's indigenous communities once existed, that is responsible for diluting the role of indigenous medicinal systems (Brown 1992). In spite of the effort of colonial governments throughout the world to undermine the indigenous systems of medicine, the system has thrived, thanks to the efforts of indigenous physicians. Contrary to the impersonal allopathic system of medicine, the indigenous systems of medicine are largely individual specific and also involve the element of "trust" owing to the familiarity or popularity of the physician in the local community.

Moreover, indigenous systems of medicine are also known to employ spiritual healing along with the therapeutic agents, besides being economical due to the non-commercial nature of indigenous medicinal systems (Pramukh & Palkumar 2006; Cavendar & Alban 2009).

Rapid assimilation of the indigenous communities into the mainstream, often catalysed by government policies and pressure from the dominant mainstream communities, has caused the decline of indigenous systems of medicine worldwide (Stephens et al. 2006). Poor recognition and erosion of cultural values have also contributed to the "lack of interest" among the new generation of indigenous communities worldwide to take up the role of herbalist. Hence it is important to learn and document the various indigenous oral pharmacopoeias along with the value accorded to them in indigenous culture. This can provide the basis for greater recognition of the indigenous medicinal systems and their custodians. This chapter deals with the various concepts, resources and other relevant information on the systems of medicine from the four communities.

Origin and Transmission of Knowledge

Knowledge about plants, animals and rituals concerned with the healing process is acquired and transmitted from one generation to another through four main processes, viz. (1) Divination, (2) Hereditary, (3) Discipleship, and (4) Kinship.

DIVINATION

Divination is the most important form of acquiring knowledge. It is the process whereby the herbalist acquires knowledge from divine forces (gods and ancestral spirits) through dreams, hallucinations (Elferink 1999) or under the influence of liquor or brew. The person who receives the knowledge is known as diviner and plays an important role in the native culture by acting as a mediator between the divine world and the mortal world and also enjoys the trust of his society (De Feo 2003; Ogundele 2007; Lefèvre 2008). The diviner is guided by divine forces who teach them the plants and animal products that can be used as medicine (Kazembe 2007). The chants associated with the healing process, best time to collect plants, administer medicine, etc. are also acquired through this process. The process of acquiring knowledge through divination is not a one-time process but originates with respect to a patient and continues throughout the healing process, as narrated by all knowledge holders interviewed. In most indigenous societies, the healers may be either diviners or herbalists (Mgbeoji 2007) but in the Koraput region, there is no separate diviner and the herbalist doubles as the diviner. Divination is not just limited to acquisition of medicinal knowledge; it also extends to aid the herbalist in the diagnosis of diseases, as known from the Bonda community and also from communities in other parts of the world (Gessler et al. 1995). It has been shown that divination is also vital to other processes such as hunting and gathering food in various regions of the world (Walters 1973). In the absence of hallucinogenic plants, the role played by hallucinogenic plants in divination as "entheogens" is taken up by liquor and brews in the societies studied. All the resource persons who had acquired knowledge through the process of divination had consumed liquor or brew during the process.

The Kondh receives his knowledge from his gods via dreams where he is shown the nature of disease, its cause and cure. While the Kondh receives the knowledge in his dream, herbalists of other communities are guided to the place where the plant occurs by the god, who points out the plant, dosage, duration, etc. to the herbalist. Under circumstances of disease progression, the process of divination repeats to show a more powerful plant. The Bonda herbalists possess gods who are known only to them or in other words, these gods are personal and "belong" to the herbalists. The Bonda attributes his healing power to his mighty god and he keeps looking for more powerful gods so that his medicinal prowess shall continue to grow. The name and identities of such gods are not revealed to others out of fear that they might be appropriated by people envious of an herbalist's success. In all the communities, the gods responsible for imparting knowledge are invoked prior to the healing process as well as thanked through rituals often involving sacrifice of a chicken and offering of liquor once the patient is healed. It should be noted that the plant knowledge acquired through the process of divination seems to follow the rule of "Doctrine of Signatures", which is discussed under a separate heading in this chapter.

HEREDITARY

The process of acquiring knowledge and its transmission from one generation to another is hereditary. The process always follows the paternal line with the son of the herbalist inheriting the knowledge. In the absence of a male heir, brother's son or sister's son receives the knowledge in order of preference. When more than one son of the family shows interest in learning, all the sons receive the father's knowledge without any bias. However, since knowledge acquiring through divination also continues side by side, it is common that the eldest learned son fares better than the younger one and hence commands better respect from the villagers.

The knowledge regarding child delivery rests with the female elder called as *gurumani* or *gurumai* (traditional birth attendant). The knowledge about plants used in abortion and childbirth as well as minor domestic rituals are transmitted through maternal lines and remains an exclusive female domain.

DISCIPLESHIP

This system can be compared to the *nungungi* system of the Australian Aboriginals (Pearn 2005) or the mainstream *gurukula* system of India where young children were identified from their childhood and trained in the native medicinal systems. However, unlike the above-mentioned systems, here the temporal extent of association between the teacher and student is less, often not more than two years.

In this method, a student can learn from a learned herbalist who is willing to teach. The student normally selects his teacher based on his inter-village and intra-village popularity. The most important criterion employed in selecting a student by his teacher is the eagerness shown by the student in learning the profession. The student will have to observe purity as dictated by his teacher. Herbalists showing affinity towards mainstream Hinduism, ask their students to abstain from consumption of meat during the period of learning that ranges from six months to a year. They will have to pay the teacher in cash or kind. The Kondh herbalists prefer to accept cash ranging from Rs. 1,000 to 1,500 while others leave it to the student to decide. Fee paid in kind includes vegetables, fowls and liquor. The teaching begins on any day considered to be auspicious by the herbalist. The Kondh herbalists consider the full moon day to be auspicious for the commencement of teaching while others choose days of their convenience. The disciple way of learning is always a masculine domain. This system of learning is an open system since any person who is willing to learn can be accommodated by any herbalist from any community.

KINSHIP

This method of acquiring knowledge is seen in the Bonda community where it is a practice for the herbalist to be accompanied by his cousin brother while administering medicine, performing rituals associated with the healing process or while collecting plants. As a result, the cousin also acquires the knowledge. This kinship continues for ever and the cousin never becomes an independent practitioner, thus maintaining the sanctity of the relationship.

Concepts and Beliefs Associated with the Healing Process

Throughout the world, healing processes employed by indigenous communities are found to be associated with different sets of concepts and beliefs. Following are the concepts and beliefs associated with the healing processes in the communities studied.

PRINCIPLES

Of all the four communities studied, only Poraja's principle behind healing is unique. The Poraja concept of medicine is unique and uses a standardised philosophy. According to that, any medicine that involves a combination of herbs should be balanced in its property of sweetness and bitterness. If there is a sweet herb used, then it has to be balanced with a bitter herb or else, the medicine may not express its complete potential.

REASONS FOR DISEASES

(1) Kondh believe that working under the sun, drinking contaminated water and evil spirits are believed to be the reasons behind any disease.

(2) According to Gadaba, evil spirits moving in the air are responsible for diseases.

(3) In Bonda system too, evil spirits are considered to be the main reason behind all kinds of diseases.

DIET RESTRICTIONS

(1) Kondh expect the patients to abstain from consuming the meat of pea fowls, rabbits, pig, goats and also from consuming creepers and climbers.

(2) According to the Poraja system of healing, pregnant women and lactating mothers should abstain from consuming buffalo meat but are free to consume cow meat. They are also advised to abstain from consuming fruits of *Terminalia bellirica* as it may destroy the foetus.

(3) In the Gadaba and Bonda systems, diet regulations are not so prominent.

DIAGNOSIS OF DISEASES

(1) In the Kondh system of medicine, diseases are diagnosed by "feeling" the palm of the patient. Persons possessed by evil spirits are diagnosed by certain symptoms such as clasping of fists and clenching of teeth.

(2) In the Poraja system, diseases are diagnosed by symptoms and divination.

(3) The Gadaba diagnose diseases by feeling the palm of the patient.

(4) The Bonda ritualist employs divination. The wrist of the patient is held by the herbalist who looks upward to the divine forces to reveal the disease.

Practices Associated with the Healing and Collection of Herbs

KONDH

(1) The Kondh do not treat a patient bitten by a *dhamna* snake (rat snake) on a Sunday, since it is believed that the snake turns poisonous on Sunday and the patient cannot be cured.

(2) In the Kondh system of healing, medicines administered for diseases inflicted by evil spirits are accompanied by specific chants.

(3) One herbalist reported that Thursday is an inauspicious day for handling medicines, since the day is devoted for divinely purposes alone and has to be avoided unless it is an emergency. However, this belief is said to be borrowed from Hinduism where Thursday is the day meant for Goddess Lokhmi (Lakshmi). The mainstream Hindus in the Koraput region consider Thursday as a day dedicated to Goddess Lakshmi and hence they prefer not to do any hard work on the day.

PORAJA

(1) The Poraja believe that medicinal herbs collected from the cremation grounds have higher vigour as they are nurtured by the ancestral spirits.

(2) The Poraja herbalists may or may not consider certain days of the week as inauspicious for collecting herbs. One herbalist reported that Saturday is an inauspicious day for collecting herbs. The same herbalist also reported that herbs have to be collected before 12.00 noon. Others believed that herbs can be collected any time. However, the herbalist has to be on empty stomach while collection of herbs.

(3) Liquor might be administered along with the medicine as it is believed to increase the potentiality of the medicine.

(4) All the medicines administered are accompanied by specific chants.

GADABA

(1) In the Gadaba community, one herbalist reported that medicines are powerful if consumed in the afternoon; others reported no preference of time.

(2) Liquor might be administered after consumption of medicine as it is believed to increase the potency of the medicine.

(3) All the medicines administered are accompanied by specific chants.

(4) Medicines can be collected from any location but herbalists may or may not consider specific days as auspicious for collection.

BONDA

(1) In the Bonda system of healing, all treatments are accompanied by specific rituals to invoke specific gods.

(2) Herbs are collected on dry days unless it is an emergency. Summer is preferred. However, a few herbalists also reported that herbs can be collected any time.

(3) Medicine is not accompanied with liquor.

Apart from the above-discussed practices, "Doctrine of Signatures (DOS)" seems to play an important role in the process of attributing medicinal plant knowledge.

Doctrine of Signatures (DOS)

Doctrine of Signatures is an ancient concept that prophesies that the creator has left clues on plants pointing to the uses that the plant can be put. According to the doctrine, plant parts shaped like a particular human organ may bear a therapeutic impact on the respective human organ and plants possessing colours that resemble a particular disease symptom may possess the ability to cure such disease. Lev (2002) defines the concept as "A theory which asserts that the way plants (and presumably animals and minerals) look, feel, taste or react suggests their medicinal application". Plants that bear vital clues pointing to potential therapeutic uses are called as "Signature Plants". Though DOS was one of the first conceived medicinal concepts on earth, it was popularised by Theophrastus Bombast von Hohenheim in his book *Supreme Mysteries of Nature* and later by Jakob Bohme through his book *Signatura Rerum* in the mid-sixteenth century (Pearce 2008). The Doctrine though believed to be redundant in today's world, is found to be in practice among Indigenous communities and sometimes even among mainstream societies. For instance, Dafni and Lev (2002) report the use of 14 plants on the basis of DOS from contemporary Israel.

In the Koraput region, plant knowledge that seems to have arisen from this concept is that generally that were arrived through divination process. *Bidens pilosa* whose seeds stick to the body of humans is used in making love potions to instigate romance in the females; *Firmiana colorata* that shows a swollen region at the point of joints of branches is used against fractures and joint pains while latex yielding plants are selected as galactagogues. Table 13 shows the properties of plants and the uses they are put to taking clues from their morphology or characteristic features as prescribed by the communities in the study area.

Apart from the above-mentioned plants, the herbalists irrespective of the community were unanimously of the view that plants that taste bitter and astringent can be used to heal dysentery and diarrhoea. This could be the rationale behind using *Bryophyllum calycinum, Gardenia latifolia* and *Syzygium cumini* for treating diarrhoea and dysentery. Though the herbalists might not use a plant, they still might ascribe DOS principles to a plant. Thus *Stereospermum colais* was identified as a tree that can be used to treat snake bites owing to its snake-like fruits. However, none of the herbalists had used this species for snake bites before.

Table 13: Signature Plants and Their Uses

Binomial	Community	Inherent Property	Uses Attributed
Bidens pilosa	Kondh, Gadaba	Seeds that stick to human beings	Love potion
Breynia retusa	Kondh, Gadaba, Bonda	Red-coloured fruits	Dysentery
Canscora diffusa	Bonda	Plant turning red when dry	Fatal disease occurring on the crown of fowls
Cissampelos pareira	Kondh	Leaves resembling urinary bladder	Urinary trouble
Cryptolepis buchananii	Kondh, Poraja, Gadaba	Latex	Galactagogue
Firmiana colorata	Poraja	Swollen joints	Fractures and joint pain
Hiptage benghalensis	Kondh	Clasping nature	Cuts
Melia azedarach	Poraja	Bitterness	Snake bite
Mimosa pudica	Bonda	Touch-me-not	Tranquilliser
Oroxylum indicum	Poraja	Fruits resembling the hood of a snake	Snake bite
Rhaphidophora pertusa	Kondh, Poraja, Bonda	Swollen nodes, Epiphytic nature	Fractures
Senecio nudicaulis	Kondh	Latex	Galactagogue

Plants in Healing

The healing process is largely of two types, viz. ritualistic and drug-based. Generally, both these methods are used in combination. The particulars of plants used by the various communities are discussed below.

KONDH

Kondh were found to use 72 plants for healing purposes including two plants that were used exclusively for rituals concerned with healing (*Maytenus hookeri* and *Paederia foetida*). Euphorbiaceae was the largest family used for medicinal purposes, followed by Asteraceae, Fabaceae and Lamiaceae (Table 14). There are 37 families from which only one species is used for healing purposes.

Table 14: Families Used by the Kondh for Medicinal Purposes (in Descending Order)

No. of Species Used for Healing	Families
9	Euphorbiaceae
7	Asteraceae, Fabaceae, Lamiaceae
5	Apocynaceae, Caesalpiniaceae, Poaceae,
4	Acanthaceae, Anacardiaceae, Apiaceae, Combretaceae, Rutaceae
3	Asclepiadaceae, Lythraceae, Malvaceae, Sterculiaceae, Rubiaceae
2	Araceae, Arecaceae, Convolvulaceae, Cucurbitaceae, Hypoxidaceae, Lauraceae, Meliaceae, Mimosaceae, Ranunculaceae, Rhamnaceae, Solanaceae, Verbenaceae
1	Acoraceae, Adiantaceae, Agavaceae, Aloaceae, Amaranthaceae, Asparagaceae, Bignoniaceae, Bombacaceae, Buxaceae, Cannabaceae, Capparaceae, Caricaceae, Celastraceae, Colchicaceae, Costaceae, Crassulaceae, Dioscoreaceae, Dracaenaceae, Gentianaceae, Hemionitidaceae, Malpighiaceae, Menispermaceae, Moraceae, Myrsinaceae, Myrtaceae, Nyctaginaceae, Oleaceae, Papaveraceae, Piperaceae, Plumbaginaceae, Polygonaceae, Sapindaceae, Sapotaceae, Scrophulariaceae, Smilacaceae, Urticaceae, Vitaceae.

Table 15. Medicinal Plants Used for Various Ailments by the Kondh

Ailment	Species
Anti-lice	*Holarrhena pubescens*
Anti-obesity	*Limnophila conferta, Rotala illecebroides*
Antiseptic	*Tridax procumbens, Ziziphus rugosa*
Asthma	*Indigofera cassioides*
Body pain	*Asparagus racemosus, Oroxylum indicum, Rhaphidophora pertusa,*
Keeping body warm	*Pongamia pinnata*
Burns	*Aloe vera*
Child delivery	*Achyranthes aspera, Artemisia japonica, Pterospermum acerifolium, Pterospermum xylocarpum*
Cold	*Terminalia chebula, Ziziphus rugosa*
Cough	*Aegle marmelos, Clematis roylei, Pogostemon benghalensis, Terminalia chebula*

Skin problems	*Amaranthus spinosus, Argemone mexicana, Momordica charantia var. muricata, Schleichera oleosa*
Deworming	*Eryngium foetidum*
Diarrhoea	*Boerhavia diffusa, Bryophyllum calycinum, Lantana camara, Piper nigrum*
Dysentery	*Alstonia scholaris, Breynia retusa, Mangifera indica, Syzygium cumini*
Emetic	*Cynodon dactylon*
Expelling insects from nostrils	*Bauhinia vahlii*
Fever	*Abelmoschus angulosus, Asparagus racemosus, Cinnamomum caudatum, Pogostemon stellatus, Sarcococca coriacea*
Fits	*Ardisia solanacea, Breynia retusa, Oroxylum indicum, Pogostemon stellatus, Thysanolaena maxima*
Galactagogue	*Cryptolepis buchananii, Senecio nudicaulis*
Hairfall control	*Datura stramonium*
Hand pain	*Polygonum chinense*
Headache	*Ardisia solanacea, Tragia involucrata*
Indigestion	*Solanum torvum*
Injuries	*Hiptage benghalensis, Litsea glutinosa*
Love potion	*Bidens pilosa*
Mental illness	*Smilax perfoliata, Solanum torvum*
Snake bite	*Ziziphus rugosa*
Sprain	*Schleichera oleosa*
Stomach ache	*Acorus calamus, Capparis olacifolia, Costus speciosus, Merremia vitifolia, Rauvolfia serpentina*
Styptic	*Stachytarpheta urticifolia*
Swellings	*Argyreia nervosa, Mimosa pudica, Thysanolaena maxima*
Tonic	*Macrotyloma uniflorum, Oroxylum indicum*
Toothache	*Zanthoxylum armatum*
Urinary disorders	*Cissampelos pareira*
Veterinary	*Justicia adhatoda, Clausena excavata, Cryptolepis buchananii,Debregeasia velutina, Euphorbia nivulia, Mucuna pruriens, Plumeria rubra, Zanthoxylum armatum*
Wounds	*Remusatia vivipara, Schleichera oleosa, Senna hirsuta*

The Kondh recognise at least 34 types of ailments and use plant species effectively to cure them. They treat not just physical ailments, but also mental ailments using species such as *Smilax perfoliata* and *Solanum torvum*. The Kondh also use eight plant species effectively for ailments in cattle. Table 15 gives the list of ailments recognised by the Kondh and the plant species used to treat them.

PORAJA

Around 81 plants used for healing purposes were recorded from the Poraja community. This includes seven plants that were used exclusively in rituals concerned with healing purposes. Euphorbiaceae is the family that has contributed the highest number of plants used for healing purpose, followed by Fabaceae and Rubiaceae (Table 16).

Table 16: Families Used by the Poraja for Medicinal Purposes (in Descending Order)

No. of Species used for healing	Families
7	Euphorbiaceae
5	Fabaceae, Rubiaceae
4	Asteraceae, Lamiaceae
3	Acanthaceae, Anacardiaceae, Apiaceae, Apocynaceae, Combretaceae
2	Araceae, Arecaceae, Asclepiadaceae, Meliaceae, Mimosaceae
1	Acoraceae, Agavaceae, Asparagaceae, Bignoniaceae, Bombacaceae, Caesalpiniaceae, Cannabaceae, Caricaceae, Celastraceae, Convolvulaceae, Costaceae, Hemionitidaceae, Lauraceae, Lythraceae, Malvaceae, Menispermaceae, Myrsinaceae, Myrtaceae, Oleaceae, Piperaceae, Plumbaginaceae, Poaceae, Ranunculaceae, Rhamnaceae

The Poraja recognise at least 32 types of ailments. They use four plant species for treating various ailments occurring in cattle, which is lesser than that of the Kondh, who use eight species to cure veterinary ailments. Table 17 gives a list of ailments recognised by the Poraja and the plant species used to treat them.

Table 17. Medicinal Plants Used for Various Ailments by the Poraja

Ailment	Species
Antiseptic	*Tridax procumbens*
Body pain	*Phyllanthus amarus, Rauvolfia serpentina, Schleichera aleosa*
Keeping body warm	*Pongamia pinnata*
Chest pain	*Rubia cordifolia*
Child delivery	*Furcraea foetida, Oroxylum indicum, Pterospermum acerifolium*
Constipation	*Argyreia nervosa, Breynia retusa, Elephantopus scaber, Eryngium foetidum*
Cough	*Cannabis sativa, Clematis roylei, Terminalia chebula*
Skin problems	*Justicia adhatoda, Aegle marmelos, Bombax ceiba, Cissampelos pareira, Lannea coromandelica, Pongamia pinnata*
Deworming	*Mucuna pruriens, Rubia cordifolia*
Dysentery	*Buchanania lanzan, Cocos nucifera, Gardenia latifolia, Syzygium cumini, Terminalia tomentosa*
Eye pain	*Ardisia solanacea, Euphorbia hirta*
Fever	*Cassia fistula, Pogostemon stellatus, Rubia cordifolia*
Fits	*Cissampelos pareira, Crotalaria pallida, Desmodium gangeticum, Rubia cordifolia, Solanum torvum*
Fractures	*Firmiana colorata, Rhaphidophora pertusa, Terminalia tomentosa, Terminalia chebula*
Galactagogue	*Carica papaya, Cryptolepis buchananii, Pedilanthus tithymaloides*
Headache	*Acacia sinuata, Andrographis paniculata, Costus speciosus, Leucas aspera*
Indigestion	*Argyreia nervosa, Peucedanum dhana var. dalzellii*
Injuries	*Litsea glutinosa, Mimosa pudica, Semecarpus anacardium,*
Malaria	*Acacia sinuata*
Measles	*Oroxylum indicum, Phyllanthus amarus*
Post partum care	*Buchanania lanzan, Indigofera cassioides, Panicum miliaceum, Phoenix acaulis*

Snake bite	*Clerodendrum serratum, Melia azedarach, Oroxylum indicum*
Scrotum pain	*Plumbago zeylanica, Rubia cordifolia*
Sprain	*Artemisia japonica, Schleichera oleosa*
Stomach ache	*Alstonia venenata, Andrographis paniculata, Asparagus racemosus, Chionanthus mala-elengi, Cissampelos pareira, Clerodendrum viscosum, Lannea coromandelica, Melia azedarach, Rauvolfia serpentina, Syzygium cumini, Woodfordia fruticosa*
Swellings	*Bombax ceiba, Euphorbia nivulia, Phyllanthus amarus, Plumeria rubra, Ziziphus jujuba*
Tonic	*Calotropis gigantea, Cassia fistula*
Toothache	*Elephantopus scaber, Mimosa pudica, Tridax procumbens, Zanthoxylum armatum*
Urinary disorders	*Thunbergia fragrans*
Veterinary	*Justicia adhatoda, Azadirachta indica, Leucas aspera, Tragia involucrata*
White spots in the tongue	*Hydrocotyle rotundifolia*
Wounds	*Guizotia abyssinica, Phyllanthus amarus, Remusatia vivipara, Stachytarpheta urticifolia, Woodfordia fruticosa*

GADABA

The Gadaba community uses 56 plants for healing purposes, including 10 plants that are used exclusively in rituals concerned with the healing process. Combretaceae and Euphorbiaceae contributed the highest number of plant species used for healing purposes (Table 18).

The Gadaba recognise 26 kinds of ailments. They also use five plant species to treat various cattle diseases which is nearly 9 per cent of the total plants used for healing purposes. Table 19 shows the various ailments recognised by the Gadaba and the plants used to cure them.

Table 18: Families Used by the Gadaba for Medicinal Purposes (in Descending Order)

No. of Species Used for Healing	Families
4.	Combretaceae, Euphorbiaceae
3.	Anacardiaceae, Apocynaceae, Arecaceae, Rutaceae
2.	Acanthaceae, Asclepiadaceae, Caesalpiniaceae, Lamiaceae, Lythraceae, Poaceae, Sterculiaceae
1.	Apiaceae, Asparagaceae, Asteraceae, Colchicaceae, Convolvulaceae, Fabaceae, Hemionitidaceae, Lauraceae, Malvaceae, Meliaceae, Mimosaceae, Ranunculaceae, Rhamnaceae, Myrsinaceae, Sapindaceae, Scrophulariaceae, Smilacaceae, Myrtaceae, Papaveraceae, Costaceae, Cucurbitaceae, Dioscoreaceae

Table 19: Medicinal Plants Used for Various Ailments by the Gadaba

Ailment	Species
Anti-Flea	*Pogostemon stellatus*
Body pain	*Schleichera oleosa*
Keeping body warm	*Pongamia pinnata*
Child delivery	*Argyreia nervosa, Pterospermum acerifolium*
Chicken Pox	*Aegle marmelos*
Constipation	*Plumeria rubra*
Cough	*Anogeissus latifolia, Barleria strigosa, Clematis roylei*
Skin problems	*Azadirachta indica, Sida cordata*
Diarrhoea	*Bauhinia vahlii*
Dysentery	*Breynia retusa*
Eye pain	*Ardisia solanacea, Calotropis gigantea, Rotala illecebroides,*
Fever	*Costus speciosus, Eleusine coracana, Mangifera indica, Pogostemon stellatus*
Galactagogue	*Cryptolepis buchananii, Phoenix acaulis,*
Headache	*Terminalia tomentosa*
Injuries	*Litsea glutinosa, Semecarpus anacardium*
Joint pain	*Diplocyclos palmatus*
Love potion	*Bidens pilosa, Mimosa pudica*

Nausea	*Diplocyclos palmatus*
Post-partum care	*Firmiana colorata, Holarrhena pubescens, Phoenix acaulis*
Snake bite	*Aristida setacea*
Sprain	*Schleichera oleosa*
Stomach ache	*Centella asiatica, Dioscorea bulbifera, Lannea coromandelica, Mimosa pudica, Pterospermum acerifolium, Rauvolfia serpentina, Syzygium cumini, Thunbergia fragrans, Woodfordia fruticosa*
Swellings	*Gloriosa superba, Murraya paniculata*
Tonic	*Mucuna pruriens*
Toothache	*Limnophila conferta, Zanthoxylum armatum*
Veterinary	*Limnophila conferta, Mucuna pruriens, Tridax procumbens, Zanthoxylum armatum*
Wounds	*Calotropis gigantea*

BONDA

The Bonda use 46 plants for healing purposes. Among these, five plants are used exclusively in rituals concerned with the healing process. Fabaceae and Lamiaceae are the largest represented families. There are 29 families that contributed one species each (Table 20).

Table 20: Families Used by the Bonda for Medicinal Purposes (in Descending Order)

No. of Species Used for Healing	Families
3	Fabaceae, Lamiaceae
2	Asclepiadaceae, Asteraceae, Hypoxidaceae, Ranunculaceae, Rubiaceae
1	Acanthaceae, Adiantaceae, Agavaceae, Amaranthaceae, Apocynaceae, Araceae, Asparagaceae, Caesalpiniaceae, Costaceae, Crassulaceae, Cucurbitaceae, Dracaenaceae, Euphorbiaceae, Gentianaceae, Lauraceae, Lythraceae, Malvaceae, Menispermaceae, Mimosaceae, Moraceae, Myrsinaceae, Papaveraceae, Piperaceae, Poaceae, Rutaceae, Smilacaceae, Solanaceae, Sterculiaceae, Vitaceae

The Bonda identify 21 types of ailments that are cured using plant species. They also use four plant species to treat veterinary health problems. Table 21 shows the list of ailments recognised by the Bonda and the plant species used to cure them.

Table 21: Medicinal Plants Used for Various Ailments by the Bonda

Ailment	Species
Abdomen pain	*Alstonia venenata*
Anti-lice	*Alstonia venenata, Ocimum canum*
Body pain	*Achyranthes aspera, Canthium dicoccum, Pogostemon benghalensis, Sida cordata*
Keeping body warm	*Pongamia pinnata*
Child delivery	*Achyranthes aspera, Breynia retusa, Mimosa pudica, Pterospermum xylocarpum*
Chicken Pox	*Aegle marmelos*
Cough	*Costus speciosus, Piper nigrum*
Skin problems	*Curculigo orchioides, C. trichocarpa*
Diarrhoea	*Cissampelos pareira, Datura stramonium, Pogostemon benghalensis*
Dysentery	*Breynia retusa*
Emetic	*Tamarindus indica*
Eye pain	*Ampelocissus tomentosa*
Fever	*Colebrookea oppositifolia*
Fractures	*Rhaphidophora pertusa*
Headache	*Vicoa auriculata*
Injuries	*Cynodon dactylon, Litsea glutinosa*
Snake bite	*Selaginella nairii*
Stomach ache	*Asparagus racemosus, Diplocyclos palmatus, Elephantopus scaber, Furcraea foetida, Hemidesmus indicus, Pogostemon benghalensis, Smilax perfoliata, Thalictrum foliolosum*
Styptic	*Argemone mexicana*
Toothache	*Breynia retusa, Crotalaria pallida*
Tranquilliser	*Mimosa pudica*
Veterinary	*Bryophyllum calycinum, Canscora diffusa, Ficus asperrima, Lagerstroemia parviflora*

The study shows that the Kondh, Poraja and Gadaba systems of healing draw 12.5, 8.6 and 7 per cent of plant species respectively from the family Euphorbiaceae. This shows the importance accorded to the family in these systems of medicine. In the case of Bonda, Fabaceae and Lamiaceae are the largest represented families. It appears that Fabaceae and Euphorbiaceae are preferred for medicinal purposes throughout the Eastern Ghats. For instance, the report of The All India Co-ordinated Research Project on Ethnobiology for Andhra Pradesh (Botanical Survey of India 1988) recorded that Fabaceae was the most preferred family for medicinal purposes (8.5 per cent), immediately followed by Euphorbiaceae (6.8 per cent). A study done on the Irula community of the Eastern Ghats in Tamil Nadu by Manjula (2004) shows that Euphorbiaceae (7.9 per cent) was the most preferred family, immediately followed by Fabaceae (7 per cent). The report of the Export-Import Bank of India (2003) too lists Fabaceae as the topmost family in Indian medicinal plant trade.

The highest number of plant species is used to treat stomach ache by all the four communities [Kondh (5), Poraja (11), Gadaba (9) and Bonda (8)]. This number goes up if diarrhoea and dysentery are also taken into account and it should be borne in mind that stomach ache is an umbrella term indicating various ailments. The Kondh use equal number of plant species for fever and stomach ache (five each). The fact that stomach-related ailments are frequent and accorded high priority in native systems of medicine has also been recorded in several other regions also. For example, Southern Spain (Rivera et al. 2005), Nigeria (Okoli et al. 2007), Assam (Sajam & Gosai 2006) and Sikkim (Pradhan & Badola 2008). This shows a general pattern that larger the incidence of a disease, higher the number of plant species used to treat that disease (Rivera et al. 2005).

A cross-cultural analysis of the plants used by the four communities for healing purposes shows that out of the total 145 plant species used for healing purpose by the communities studied, 36 species are used for the same purpose by more than one community. Two species are used for the same purpose by all the four communities while 12 species are used for the same purpose by three communities. Our observation shows that *Litsea glutinosa* and *Pongamia pinnata* are used for the same purpose by all the communities inhabiting the Koraput region and hence is a common knowledge. *L. glutinosa* is commonly used by all the communities in the Koraput region and also by the communities in the Eastern Ghats of Andhra

Pradesh for treating fractures and cuts. Communities in the Koraput region identify the plant by the local name *tangia achina* while the communities in Andhra Pradesh call it as *pangia achina* (Botanical Survey of India 1988; Rao and Henry 1996). This clearly shows that the knowledge is common for communities inhabiting a wider range of area. Similarly, the use of the oil from the seeds of *P. pinnata* to keep the body warm during winter is also a public domain knowledge, irrespective of the community and individual. *T. bellirica* and *T. chebula* are identified by the communities as part of the *triphala* of Ayurveda. However, they are not clear about the uses attributed to it. This knowledge has diffused into the community through the various training programmes imparted in Āyurveda by the non-governmental bodies operating in the area.

The presence of same use value for the same species in more than one community also indicates a natural validation and suggests promise of cure for the ailment. However, the fact that just 36 species out of 145 species are used for the same purpose by more than two communities and a mere two species is used for the same purpose by all the four communities show that the knowledge is largely not common to all the communities. Within the community, the knowledge again exists in private domain, i.e. in the possession of specialist healers who put it to non-commercial use. Table 22 shows the list of plants used for the same purpose by more than one community.

Table 22: Plants Used by More than One Communities for the Same Purpose

Binomial	Purpose	Communities
Achyranthes aspera	Childbirth	Kondh, Bonda
Justicia adhatoda	Skin diseases in cattle	Kondh, Poraja
Aegle marmelos	Rituals Chicken pox	Kondh, Poraja Gadaba, Bonda
Ardisia solanacea	Eye pain	Poraja and Gadaba
Argemone mexicana	Rituals	Kondh, Gadaba and Bonda
Asparagus racemosus	Stomach pain Rituals	Poraja and Bonda Gadaba and Bonda
Bidens pilosa	Love potion	Kondh and Gadaba
Breynia retusa	Dysentery	Kondh, Gadaba and Bonda
Cassia fistula	Rituals	Poraja and Gadaba

Clematis roylei	Cough	Kondh, Poraja and Gadaba
Cryptolepis buchananii	Galactagogue	Kondh, Poraja and Gadaba
Hemionitis arifolia	Rituals	Poraja and Gadaba
Lannea coromandelica	Stomach ache	Poraja and Gadaba
Litsea glutinosa	Injuries	All
Maytenus hookeri	Rituals	Kondh, Poraja
Mucuna pruriens	Anthelmentic in cattle	Kondh, Poraja and Gadaba
Paederia foetida	Rituals	Kondh, Poraja and Bonda
Phoenix acaulis	Post-partum care	Poraja and Gadaba
Phyllanthus emblica	Rituals	Poraja and Gadaba
Pogostemon benghalensis	Rituals	Bonda and Gadaba
Pogostemon stellatus	Febrifuge	Kondh, Poraja and Gadaba
Pongamia pinnata	Keep body warm	All
Pterospermum acerifolium	Childbirth	Kondh, Poraja and Gadaba
Pterospermum xylocarpum	Childbirth	Kondh, Bonda
Rauvolfia serpentina	Stomach ache	Kondh, Poraja and Gadaba
Remusatia vivipara	Wounds	Kondh and Poraja
Rhaphidophora pertusa	Fractures	Poraja and Bonda
Schleichera oleosa	Sprains	Kondh, Poraja and Bonda
Semecarpus anacardium	Minor injuries	Poraja and Gadaba
Smilax perfoliata	Rituals	Poraja, Gadaba and Bonda
Syzygium cumini	Dysentery Stomach ache	Kondh and Poraja Poraja and Gadaba
Terminalia bellirica	Part of the Ayurvedic *triphala*	Kondh, Poraja and Gadaba
Terminalia chebula	Part of the Ayurvedic *triphala* Cough	Kondh, Poraja and Gadaba Kondh and Poraja
Tridax procumbens	Antiseptic	Kondh and Poraja
Woodfordia fruticosa	Stomach ache	Poraja and Gadaba
Zanthoxylum armatum	Toothache	Kondh, Poraja and Gadaba

Plants Used in Ritualistic Healing

Table 23: Plant Species Used in Healing Rituals

Community	No. of Species	Binomial
Kondh	4	*Argemone mexicana, Maytenus hookeri**, *Paederia foetida**, *Pogostemon stellatus**
Poraja	15	*Acorus calamus**, *Aegle marmelos, Cassia fistula, Clerodendrum serratum, Cocos nucifera, Hemionitis arifolia**, *Leucas aspera, Maytenus hookeri**, *Mucuna pruriens, Paederia foetida**, *Phyllanthus emblica**, *Rubia cordifolia, Smilax perfoliata**, *Solanum torvum, Urena lobata**
Gadaba	10	*Argemone mexicana**, *Asparagus racemosus**, *Bridelia retusa**, *Cassia fistula**, *Hemionitis arifolia**, *Phyllanthus emblica**, *Pogostemon benghalensis**, *Ricinus communis**, *Smilax perfoliata**, *Ziziphus rugosa**
Bonda	14	*Adiantum philippense**, *Andrographis paniculata**, *Ardisia solanacea**, *Argemone mexicana, Asparagus racemosus, Clematis roylei**, *Cryptolepis buchananii**, *Desmodium triquetrum**, *Dracaena terniflora**, *Ficus asperrima, Paederia foetida**, *Pogostemon benghalensis, Pterospermum xylocarpum, Smilax perfoliata*

Note: * denotes plants used exclusively for healing rituals.

The Kondh use the least number of plant species for ritual purposes and the Bonda use more number of species for healing purposes (Table 23). The spiritual nature of the Bonda and their reverence for supernatural forces are probably the reasons for their greater dependency on ritualistic healing. The number of plants used as drugs for therapeutic purpose by a community is inversely proportional to the number of species used exclusively in healing rituals by that community (*fig*. 2).

Kondh use 4.16 per cent of the total plants used for healing purposes, in healing rituals. Poraja use 18.5 per cent species, Gadaba use 17.85 per cent and Bonda use 30.4 per cent. Except the Gadaba where the plants used in rituals are not used as drugs, all other communities use a few plants both for rituals as well as for drugs. Kondh use one species for both rituals and as drug, Poraja use seven species and Bonda use six species.

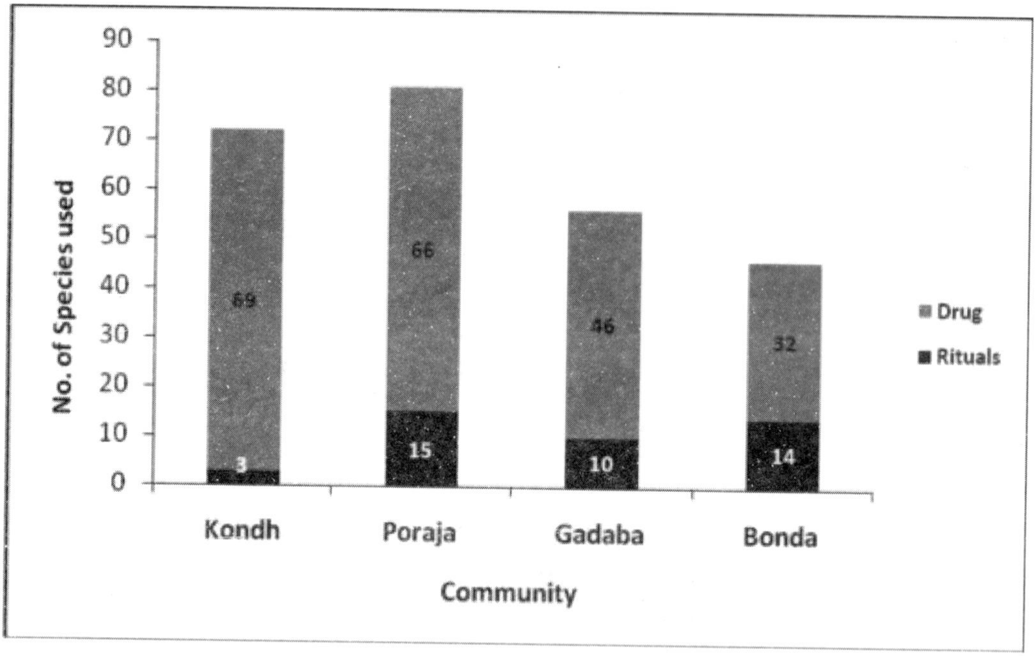

fig. **2: Comparative Chart Showing the Total Number of Plants Used for Healing Purposes and the Plants Used for Ritual Purposes**

The Bonda are highly spiritual. Their medicinal system lays emphasis on spiritual healing and thus plant-based drugs are only secondary or supplementary means. This stands in contrast with the idea of the herbalists from other communities that drugs are the main medium for curing diseases and invocation of spirits is only a measure of "enhancing" the quality of the drug. This can be further understood by the common belief among the Bonda herbalists that any plant when blessed by the appropriate spirit, can act as a healing agent. The study shows that the Kondh and Poraja are following a more systematic and evolved plant-based medicinal system while the Gadaba and Bonda follow their age-old system of healing rooted in their religion.

Exotic Plants Used for Healing

The communities use 12 exotic plant species for healing purposes (Table 24).

Table 24. Exotic Plants Used for Healing Purposes by the Communities

Binomial	Nativity	Uses			
		Kondh	Poraja	Gadaba	Bonda
Aloe vera	Macronesia	Burns	-	-	-
Amaranthus spinosus	Tropical American	Boils	-	-	-
Argemone mexicana	Tropical American	Skin diseases	-	-	Injuries, fever
Bidens pilosa	Tropical American	-	-	Love	- potion
Furcraea foetida	Tropical American	-	Ease delivery	-	Stomach ache
Lantana camara	Tropical American	Hardening stools	-	-	-
Mimosa pudica	Tropical American	Swellings	Injuries, gum ache	Love potion	Child delivery, induce sleep
Phyllanthus amarus	Tropical American	-	Measles, wounds, swellings	-	-
Plumeria rubra	Tropical American	Veterinary	Swellings	Laxative	-
Senna hirsuta	Tropical American	Wounds	-	-	-
Solanum torvum	Tropical American	Mental illness, stomach problems	Fits	-	
Tridax procumbens	Tropical American	Antiseptic	Antiseptic, gum ache	Veterinary	-

Kondh use the highest number of exotic plant species for medicinal purposes, followed by Poraja with nine species. Most of the species are Tropical American in origin. Though all these species are naturalised and commonly found in the Koraput region, the use patterns vary and no species is found to be used for the same purpose by two communities. This shows that the knowledge on these plants has evolved *sui generis* and belong to these communities.

6

Forests and Non-Timber Forest Products

When communities are empowered with greater rights and responsibilities, forest conservation and rural livelihoods tend to improve.
— The World Bank 2006

Forests

THE forests found in the study area can be classified into two, viz. (i) Natural forests, and (ii) regenerated forests. Natural forests can be further classified into two sub-groups, viz. (a) secondary natural forests, and (b) community protected natural forests. Other than these, plantations are also found.

(i) Natural forests are those growing spontaneously with no or little support from the community.

  (a) Secondary natural forests are forests allowed to regenerate after a period of shifting cultivation. Such forests are not protected consciously by the communities and are found in the Bonda Hills.

  (b) Community-protected natural forests are forests that were not regenerated, according to the memory of the present generation of the people. Sacred groves and cremation grounds also fall into this category (*figs.* 62 and 63). These are part of original forest patches that once existed and were protected by the communities. Examples of such forests are Chikalmari, Panasput and Gullel.

(ii) Regenerated community-protected forests are those forests regenerated in areas where shifting cultivation once existed. Later the area was spared from shifting cultivation and the forest was allowed to regenerate. Examples of such forests are Putisil, Kondh-pungar, Khojuriguda, Poraja-pungar, Pipalguda, Chompakanda, Billaput and Dokriguda.

(iii) Plantations occur in villages such as Bonduguda, where the hills were afforested with species such as *Acacia auriculiformis, Eucalyptus spp.* or *Grewelia robusta* by the forest section of soil conservation department. Such plantations are meant to prevent soil erosion and offer conducive conditions for the native species to survive beneath.

COMMUNITY MANAGED FORESTS IN THE STUDY AREA

Putisil: The forest of Putisil is a regenerated one. The forest has been protected by the whole village. The forest is an example of organised forest management, introduced by IRDWSI. The community had to undertake serious afforestation programmes and protection methods due to the mini-hydropower plant that came up with the help of IRDWSI. The dominant species are *Acacia sinuata, Hiptage benghalensis, Artocarpus heterophyllus, Chionanthus mala-elengi* and *Wendlandia glabrata.* Animals found include wild boar, jackals, bears and spotted deer.

Kondh-pungar: Kondh-pungar shares the same hill with the neighbouring Poraja-pungar. There is a space of 5 ft between the forests of both these villages, which serves as boundaries. The forest has been protected for the past 30 years. The whole village is involved in the protection of the forest and being a small village, forest related offences do not occur. The forest is prone to grazing and is found heavily infested with *Lantana camara. Buchanania lanzan, Phyllanthus emblica, Terminalia bellirica* and *Terminalia tomentosa* are the dominant species.

Chikalmari: The forest of Chikalmari is also called Kachelumari and is a natural forest protected by the villagers. The forest is protected by a community-appointed forest guard who is paid 6 kg of any grain every month by each family of the village. There are no forest-related offences committed by the villagers. However, the village forest is under pressure from the neighbouring villagers. The dominant species are *Cinnamomum caudatum, Maytenus hookeri* and *Neolitsea foliosa.*

Bonduguda: The forest of Bonduguda is a plantation of *Acacia auriculiformis*, undertaken by the forest department. However, the acacia trees are now being systematically cleared by the villagers, thus providing opportunity for the native trees growing underneath to be exposed and to get established.

Panasput: According to the memory of the present generation the forest of Panasput is a natural forest which was offered protection. The forest was once protected by specially appointed forest guards and now it is protected by the whole village. Trees are permitted to be cut only with the prior permission of the village council and the weapons of the violators are seized and a penalty is imposed depending upon the cache. The dominant species are *Mangifera indica, Wendlandia glabrata, Chionanthus mala-elengi* and *Neolitsea foliosa*. Animals found include bears and spotted deer.

Gullel: The forest of Gullel is a natural forest protected by the community from 1987. The forest shows dense vegetation and is a home to animals such as bears and sambar deer. The dominant tree species is *Mangifera indica*. There is no penalty imposed on violators who cut trees without the permission of the village council, but the wood is seized, burnt and the tools confiscated.

Khojuriguda: Khojuriguda owns a community-protected forest that was allowed to regenerate about a decade back. The whole village is involved in protecting the forest and being a small village, no forest-related offence takes place.

Poraja-pungar: Poraja-pungar shares the same hill with the neighbouring Kondh-pungar. Hence the forests of these villages lie on the same hill adjacently, separated by a gap of 5 ft, cleared by the villagers of Kondh-pungar for marking purpose. The forest has been protected for the past 30 years. There are two forest guards who are paid 3 kg of rice and Rs. 10 by each family every month. The forest is prone to grazing and is found heavily infested with *Lantana camara. Buchanania lanzan, Phyllanthus emblica, Terminalia bellirica* and *Terminalia tomentosa* are the dominant species.

Pipalguda: The forest of Pipalguda is maintained by Gangagudi Mahila Mandal, a women's self-help group. The villagers are supposed to get prior permission from the *mandal* before cutting trees from the forest. The *mandal* is empowered to seize the tools and impose a penalty of Rs. 40-50 in case of violations. While harvesting timber, branches of big trees are cut and in the absence of big trees, small trees would be cut. Trees of *Mangifera indica, Artocarpus heterophyllus* and *Syzygium cumini* are not cut. The forest has been protected for the past 15 years and hence shows luxuriant growth.

Chompakanda: Their community-protected forest lies on the mountain of Soliababuli which was allowed to regenerate naturally in 1982, after the intervention of IRDWSI. The lower slopes of the mountain were planted with species such as *Syzygium cumini* and *Mangifera indica*. There is a forest guard appointed by the village. He is paid around 3 kg of rice, 3 kg of foxtail millet (*suan*) and 3 kg of pearl millet by each family every month. There is no penalty imposed upon people who cut trees without permission. However, they are given corporal punishment.

Billaput: The forest of Billaput was allowed to regenerate around eight years back and hence is sparsely vegetated. The forest is protected by the village youth committee called as *Jeevak Dal*.

Dokriguda: The forest of Dokriguda was allowed to regenerate around 15 years back and hence shows dense vegetation. The forest guard is paid 9 kg each of rice, pearl millet and foxtail millet every month and a share of the vegetable cultivated annually by each family. *Diospyros sylvatica, Mangifera indica* and *Semecarpus anacardium* are the dominant species. The lower part of the mountain is afforested with *Anacardium occidentale*.

Bonda Hills: The concept of community-protected forests does not exist in the Bonda hills. The Bonda settlements are surrounded by secondary forests allowed to regenerate after the period of shifting cultivation is over. The forest is common to all the members of Bonda community and hence NTFP can be collected irrespective of the village which one hails from.

Community-based conservation efforts gained prominence since 1980s (except Bonda) possibly due to two reasons: The concept of Joint Forest Management (JFM) was introduced during this period and IRDWSI popularised and strengthened the concept of community forestry that is known to have existed already in the villages belonging to the "backward class" communities such as the Mali. It could be said that both these factors had a synergistic effect. What is understood is that Community Protected Forests existed even before the introduction of JFM and they were only strengthened later by the JFM. The JFM scheme was introduced in Odisha in the form of Village Forest Protection Committees (VFPC) in 1988 but the term JFM was first used only by 1993 (Sarap 2004). The organised conservation efforts involving the village committees are mostly influenced by JFM while conservation through community forest guards seems to be influenced by IRDWSI.

Non-Timber Forest Produce

Anthropogenic biomes played a crucial role in developing a knowledge base on the bioresources (Ellis & Ramankutty 2008). This knowledge base is key to the survival and well-being of the forest-dwelling communities. Subsistence of such communities depends upon innumerable services and products derived from the forests. The produce that is collected from the forests, other than wood for subsistence and livelihood are termed as NTFPs. A variety of NTFPs such as medicines, oils, waxes, honey, fruits, tubers, cordage, spices, etc. are collected from the forests in India. According to the All India Coordinated Research Project on Ethnobiology (Pushpangadan 1984), 7500 wild plant species are used by the tribals for medicinal purposes, 3900 species as wild edibles, 525 species for fibre and cordage, 400 species as fodder and 300 species as piscicides and pesticides. NTFPs have contributed immensely to the Tribal economy as well as to the country's economy. For instance, the export value of NTFPs for Thailand in 1987 was found to be US $ 32 billion. In the same year, the value of NTFPs exported from Indonesia stood at US $ 238 million (De Beer & McDermott 1989). NTFPs are gaining prominence today due to their potential to aid economic growth with nil or little damage to the forest from which they are harvested.

In a study conducted in the Angul and Keonjhar districts of Odisha, it was found that 95 per cent of the households got at least some cash income from NTFPs and NTFP-based products, which contributed, on average of around 19 per cent of total cash income for the community (Mahapatra et al. 2005). In the study area too, the tribal communities are dependent on the forest resources to a great extent for their daily needs. The NTFPs are found to contribute more to the family in kind than monetarily, as understood from the lesser quantity of products traded commercially. Table 25 gives an overview of NTFPs collected in the study area. Except the Bonda community, data is presented village wise since the nature of village forest varies with the village and the villagers are mostly permitted to collect NTFPs only from their own village forest, except during circumstances when the market demand for NTFPs is low. But in the Bonda community, the villagers are free to collect NTFPs from any forest and the nature of forest is also the same all over the Bonda hills and hence data from the Bonda community is presented as a single entity.

The study also reports NTFPs such as the grubs of *Rhyncophorus ferrugineus* (*fig.* 70), red ants and winged termites not considered as NTFPs in Odisha till now. Medicinal plants harvested by the communities are not included in this chapter for convenience sake and are discussed in separate chapters. The various categories of NTFPs collected in the study area are provided below.

NTFPs COLLECTED

The following kinds of NTFPs are collected by the communities.

Wild Edibles

Bauhinia vahlii, Dioscorea bulbifera, Dioscorea glabra, Dioscorea oppositifolia, Dioscorea pentaphylla, Phoenix acaulis, Syzygium cumini, Mangifera indica (*fig.* 71), *Bauhinia vahlii, Bauhinia variegata, Phyllanthus emblica, Terminalia chebula, Terminalia bellirica, Buchanania lanzan, Zizyphus rugosa, Rubus ellipticus* (*fig.* 66), *R. niveus, Acacia sinuata, Artocarpus heterophyllus, Casearia graveolens, Indigofera cassioides, Momordica charantia var. muricata, Caryota urens, Rhyncophorus ferrugineus,* Honey, Red ants and Winged termites.

Liquor Distillation

Species such as *Artocarpus heterophyllus, Mangifera indica, Caryota urens* and *Madhuca longifolia* are collected for liquor distillation by the Bonda alone (*fig.* 65).

Cordage

Bauhinia vahlii, Fleurya interrupta, Ichnocarpus frutescens (*fig.* 69), *Galactia tenuiflora, Helicteres isora, Hiptage benghalensis* and *Caryota urens.*

Household Articles

Bauhinia vahlii (Meal Plates; *figs.* 74 and 76), *Bambusa bambos* (mats, winnow, grain storer and fish trap; *figs.* 68 and 75), *Thysanolaena maxima* and *Sida acuta* (*fig.* 72).

Personal Hygiene

Lannea coromandelica (toothbrush).

Decorative

Michelia champaca, Firmiana colorata, Phoenix sylvestris, Abutilon indicum, Plumeria rubra, Phoenix acaulis and *Caryota urens.*

Thatching

Themeda triandra (*palda*), *Pennisetum hohenackeri* (*somp piri*) and *Cymbopogon gidarba* (*singani piri*) are the three species that were once used for thatching the huts until the tiled roofs replaced the thatched roofs.

Oil Extraction

Species such as *Pongamia pinnata* (*fig.* 64), *Casearia graveolens, Ricinus communis* and *Michelia champaca* are collected for extracting oil. Of these, *P. pinnata* is mostly collected from non-forest areas since not many trees are available in the wild. *M. Champaca* is collected in smaller quantities by Bonda for extracting mosquito repellent oil.

Table 25 gives an overview of the number of NTFPs collected in the study area and the purpose of collection. The plant species that were once used for thatching purposes are not included since they are not collected anymore.

COMMON SPECIES WITH POTENTIAL TO BE TRADED

Of all the species collected, a total of 15 species were found to possess extensive trade value. They are *Thysanolaena maxima* (inflorescence), *Bauhinia vahlii* (leaves),

Table 25: NTFPs Collected and the Nature of Collection

Village	Species Collected	Products Collected	Plant Products	Animal Products	Products for Lean Season Consump-tion	Products Traded	Products with Potential for Trade
Putisil	32	34	32	2	6	2	12
Kondh-Pungar	16	17	16	1	4	Nil	9
Chikalmari	21	23	22	1	1	2	1
Bonduguda	24	26	24	2	4	Nil	9
Poraja-pungar	29	32	30	2	4	Nil	11
Khojuriguda	25	28	27	1	3	1	10

Panasput	31	33	31	2	1	5	9
Gullel	26	29	27	2	1	6	6
Chompakanda	19	20	20	Nil	4	Nil	9
Billaput	18	21	19	2	3	2	6
Dokriguda	17	19	18	1	2	Nil	3
Pipalguda	27	28	26	2	5	5	7
Bondaghat	33	37	33	4	5	11	4

Syzygium cumini, Mangifera indica (fruits), *Phyllanthus emblica, Terminalia bellirica, T. chebula, Buchanania lanzan, Musa balbisiana* (leaves), *Acacia sinuata, Artocarpus heterophyllus* (fruits), *Butea monosperma* (leaves), *Bixa orellana* (seeds), *Pongamia pinnata* and *Mallotus philipensis.*

Putisil was found to collect the highest number of NTFPs immediately followed by the villages of Bondaghat. The reason for Putisil ranking high is its organised system of forest management starting with the setting up of the mini-hydropower project. The predominantly agriculture-dependent Kondh-pungar naturally ranks the least in the number of NTFP species collected. Kondh-pungar is a comparatively small hamlet, surrounded by agricultural fields. This village has taken up modern agriculture quite a long time back. With hybrid varieties and modern agricultural inputs, they are able to produce surplus tradeable agricultural commodities making them less dependent on NTFPs. Kondh-pungar and Poraja-pungar are nearby villages, sharing different forest areas of the same mountain. But the villagers of Poraja-pungar were found to collect 29 species of NTFPs. The community-protected forest of Billaput is of recent origin and thus offers little scope for NTFP collection. This explains the reason for its lower number of NTFP collection. However, this doesn't seem to be the general trend since Dokriguda and Chompakanda, two other villages with low NTFP turnout are found to possess good community-protected forests standing for more than 18 years.

THE ROLE OF NGOs AND GOVERNMENT SECTOR IN TRADE AND MANAGEMENT OF NTFPs

Till the 1970s, the tribal communities of Koraput region were known to barter their NTFP products with the mainstream communities for products such as salt, utensils and ornaments that could not be obtained/produced from their ecosystem. The barter system slowly started paving way to cash-based trading and today of the four communities studied, the Bonda continue to barter their goods while the others rarely barter. The Bonda barter *olsi* (*Guizotia abyssinica*), a non-NTFP for salt. They also barter NTFPs for utensils such as bamboo mats, baskets, threads and beads to make ornaments. The decline of the barter system and the low income obtained from NTFPs slowly led to the decline of NTFP trade.

The study shows that there are at least 14 species that possess greater trade potential. Under-utilisation of these species is due to various barriers that exist on the ground such as

1. The presence of middlemen,
2. Lesser earning, and
3. Lack of access to market.

The middlemen are from the mainstream communities and are known to exploit the ignorant tribal people. They are known to undervalue the produces as well as create an artificial local market depression, thanks to the nexus among the fellow middlemen. For instance, a comparison of the data shows that the Poraja of Panasput sold seed kernels of *Buchanania lanzan* at Rs. 10/kg while the Poraja of Gullel sold the same for around Rs. 17-20/kg during 2004-05. With both the villages situated at a distance of less than 6 km, the difference in price is not justifiable. Similarly in the same year, the Kondh of Chikalmari had sold fruits of *Terminalia bellirica* and *T. chebula* for Rs. 12 and 10 respectively. However in the same year, the Poraja of Panasput had sold the same for Rs. 2 and 10 per kg respectively. The above observation clearly shows that the different middlemen had offered prices of extremely varying nature. Receiving a price of Rs. 2/kg of *T. bellirica* in no way could have been remunerative for the people of Panasput. This can be comparable to the results of the study done by Choudhury (2007) in the Kirkicha Watershed in Kalahandi district of Odisha where tamarind sold to the middlemen at Rs. 3/kg would fetch at least Rs. 12/kg in the local market just 40 km away from the village. Choudhury also points out that the tribals incur a net loss of around 70 per cent in the total price of the NTFP when sold to a middleman. The presence of middlemen does not permit the people to obtain maximum price for their produce. But there are very few alternatives available for the people too.

Alternate marketing mechanisms such as co-operative federations and societies set up by NGOs failed to deliver justice due to minimal support from the government sector. However, with enough support from the government, these mechanisms could help the people in a great way. For example, the data for broom marketing undertaken by the Deomali Jhimkiguda Mahila Society, a co-operative women's federation involved in the collection and marketing of NTFPs show that the village Bonduguda with less *Thysanolaena maxima* available in their forest, had produced 3,377 processed brooms during 2001-02 and 1576 processed brooms during 2002-03 (IRDWSI unpublished data). This could have been possible only because the villagers were able to venture into nearby forests other than their own and carry out collection. However, the same village reported no broom collected during 2004-05 because the society was not able to market the product of the previous years and hence had stopped procuring broom by then. The reasons were high inter-state taxes that prevented the sale of the broom outside the state and also the flooding of the market with cheaper and fancy nylon brooms. It should be noted here that until the year 2000, processing of hill brooms can only be done by TDCC which was the lease-holder and its traders (Saxena 2003). This meant that the Tribal communities could sell only the unprocessed brooms to Tribal Development Co-operative Corporation (TDCC) and the marketing of processed brooms by the various co-operative societies was possible only after the year 2000.

According to the Odisha NTFP resolution 2000, NTFPs except nationalised NTFPs such as *kendu, sal* and bamboo can be procured by the *gram panchayats* (Sarap 2004). One commendable feature of these resolutions is that there would be a minimum price fixed by the district collector in consultation with the local bodies, forest department, etc. In the absence of this mechanism, the contractor-political party nexus would have taken advantage of the tribal communities. However, considering the fact that this study was done three years after these resolutions were passed, it could be said that these resolutions failed to impact NTFP trade in the study area to a visible extent. But the resolutions seemed to have their impact in other parts of the state. In the Mandibisi village of Rayagada district, a harvest of 10 kg of unprocessed broom fetched a mere Rs. 25 which had increased to Rs. 50, post-2000 resolution. In the same village, the per quintal wholesale price of broom leaped from Rs. 200 in 1991 to Rs. 3000 in 2002 (Mahapatra 2004).

NTFPs as Source of Livelihood Support

NTFPs contribute to the village economy both in terms of cash economy as well as in terms of subsistence. Sometimes, the currency brought in by the NTFPs could be less than the actual contribution of the NTFPs to the village economy. This is clearly observed in the study area where the number of NTFPs collected for domestic consumption is much higher than the number of NTFPs collected for both trade and domestic consumption. *Fig.* 3 illustrates the percentage of NTFPs collected for exclusively domestic consumption purpose.

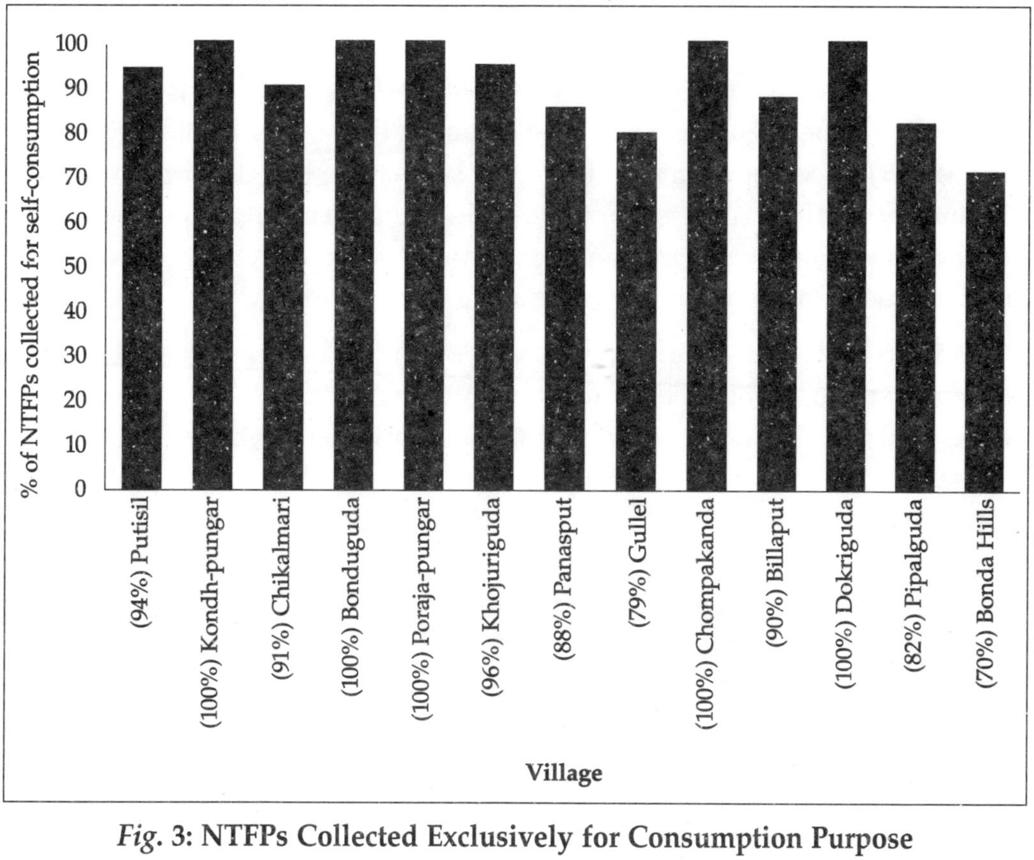

Fig. **3: NTFPs Collected Exclusively for Consumption Purpose**

A majority of NTFPs are collected for domestic consumption purpose which suggests that the communities are forest dependent. This situation is found to be similar in the Visakhapatnam, Chittoor and Adilabad districts of Andhra Pradesh in a study conducted by TERI in 2004. According to the study, of the total 37 NTFP species collected eight species were collected exclusively for self-consumption purposes and 12 species exclusively for commercial purposes. The rest 18 species were utilised locally for self-consumption purposes as well as sold locally. Thus it could be said that as many as 26 NTFP species met the daily needs of the local villagers, indicating a high degree of forest dependency. Similarly, the results of a global study undertaken by Belcher et al. in 2003 at 61 localities from Africa, Asia and Latin America, shows that the major contribution from NTFPs towards the communities were in the form of subsistence goods and their contribution in terms of cash was very low. The study also says that the villagers involved in harvesting NTFPs used a higher number of NTFPs for their own consumption than their neighbours. Hence it could be said that viewing NTFPs on the basis of their cash contribution alone might not give a clear picture of their role in the society and the fact that NTFPs contribute more in terms of subsistence is a global phenomenon. The number of NTFPs collected for trade could change depending upon the market demand as well as the interventions of agencies such as NGOs, self-help groups and forest department as well as government policies.

Table 25 also shows the quantity of NTFPs collected during lean seasons. Lean seasons are periods when there are not many agricultural or labour activities that happen and the NTFPs act as sources of support during such seasons. The common lean season is March-April when the *Choith porv* festival happens. Traditionally, this festival happened for one full month and with little agriculture produce available and the entire human resources of the village is devoted towards the game hunt. It is the harvest of NTFPs such as toddy from *Caryota urens,* the grub of *Rhyncophorus ferrugineus,* fruits of *Mangifera indica* and the meristem of young *Phoenix acaulis* that serve as lean season support mechanism. When the men folk went on hunt, the women go on collecting NTFPs and processing of Mango fruits into pickles meant for domestic consumption. On an average, an adult male member of the tribe consumes around 4 litres of toddy which provides vital calories. Though the duration of festival has come down to 7-14 days in the recent times, these NTFPs continue to provide support during the lean period. Moreover, clearing of forests for shifting cultivation also happens during March-April and the people involved in clearing depend upon toddy and the meristem of *P. acaulis* for food. In the Bonda community, liquor distilled from the fruits of *Mangifera indica* and

Madhuca longifolia provides vital support during *Choith*. The trade in NTFPs though possessing immense potential could have declined, but the NTFPs continue to provide vital support for the community as lean season support mechanism as well as fulfil their domestic needs.

Besides the above-discussed contributions of NTFPs, the study also shows that there are around 27 wild edible NTFPs collected which includes the grub of *Rhyncophorus ferrugineus*, honey, red ants and winged termites. This shows the diversity of the food resources available to the community. Having a wide food base ensures diverse nutrients and also ensures food security because the community is not compelled to depend on a few food resources. A report by Ghosh (2002) shows that even the *Rubus ellipticus* which is otherwise omitted from the NTFP list owing to its collection in lesser quantities plays a significant role in offering nutritional security (Vitamin C:38 mg/100g and 22 mg of Ca/100g). Sources from the study area say that they could survive short duration of famines by consuming porridge prepared out of dried mango kernels.

HARVEST PATTERNS

Sustainable management and development of NTFP resources is important to ensure the sustenance of the population that is traditionally dependent on them (Silori et al. 2005). Traditional sustainable harvesting techniques have not been recorded from the study area. Practices such as leaving a tuber behind whenever *Dioscorea spp.* are harvested have been introduced by the intervening agencies such as NGOs. The communities of the study area acknowledge the dwindling quantities of tubers available for harvest though the quality of forest improves as time progresses due to community forestry. This is due to the increase in population which in turn increases the demand for wild food resources availability which largely remains the same. NTFP harvest is mostly a family affair and it is the women and children who are mostly involved in the collection. NTFPs required for community feasts and festivals such as leaves of *Musa balbisiana* and *Bauhinia vahlii* are collected by the whole village and under normal circumstances, each family of the village will be allotted a quota in the form of number of bundles to be collected. This practice serves to strengthen the community bond.

The forests and their NTFPs tend to contribute more in terms of subsistence than cash. However, the cash contribution too can be increased substantially by taking various corrective measures in the ground. Some of the measures recommended to increase the contribution of NTFPs for both subsistence and trade are given below.

1. The state government should extend compensation to the communities involved in protection of forests. This could be done by taking into consideration the cost that would have otherwise been borne by the government for conserving such areas through the forest department. The payment should also effectively consider the ecosystem services rendered by the Community Protected Forests (CPF). The CPFs have a greater role to play in the wake of Odisha's fast depleting forest cover. Saxena (2003b) points out that in the period of 1980-95, Odisha had lost 9.4 per cent of its dense forest cover and at the same period, India had improved the overall dense forest cover by around 1.6 per cent.
2. The Odisha Forest Department should be provided with adequate personnel. In many cases, it was noted that while the community was effective in controlling members of their own from committing forest-related violations, they were helpless when it came to violations by other communities. The forest department should assist the community under such circumstances. Moreover the Odisha state forest department is also known to be highly corrupted (Agrawal 1995) which has led to the alienation of tribal communities from the government. Hence it is imperative for the forest department to maintain a clean record for earning the goodwill of the tribal communities.
3. The nutritional contribution of all the wild food harvested in the region has to be studied. This could give a clear idea of the contribution of NTFPs to nutritional security and also would give vital inputs for policy development.

7

Agro-biodiversity

> And as they came to live close to their fields, some men ceased to be hunters and tenders of flocks. They became farmers. Except that it was almost certainly not men who *developed this first agriculture, but rather women.*
>
> — Trager 1982

TODAY'S agro-biodiversity is based on the traditional wisdom associated with selection and breeding of crops. The existence of traditional agricultural knowledge is vital for the survival of agro-biodiversity. It should be noted here that traditional knowledge is not found as a free-floating entity but embedded in culture. Hence areas of high cultural diversity are known to possess high agro-biodiversity (Nautiyal et al. 2008; Maheshwari 1997). The Jeypore tract of the study area is known for its diversity of tribal population and also for its diversity of traditional rice varieties. In fact, the Jeypore tract is considered as one of the secondary centres of origin of rice (Patra & Dhua 2003).

The traditional method of selection and breeding was slowly sidelined by the Mendelian method of hybridisation. However, both the Mendelian method of hybridisation and the modern techniques such as genetic engineering depend upon the agro-biodiversity provided by indigenous farmers and tribal communities for the basic genetic material (Brush 2005). Bioprospecting of traditional crops for "desired" genes is an important step in genetic engineering of crop plants. For instance, a highly successful IR-72, a modern variety of rice was possible only by cross-breeding 22 landraces from India, Indonesia, China, Philippines, Vietnam, Thailand and Malaysia (Table 26).

Table 26: Pedigree of IR-72

Country	Landraces Procured
India	*Oryza nivara, arikarai, eravapandi, gowdalu, kitchili samba, latisail, mudgo, thekkan, vellaikar*, and two unknown varieties
Indonesia	*Benong*
Malaysia	*Seroup besai*
Vietnam	*Tetep*
Thailand	*Germ pai*
Philippines	*Marong paroc, sinawpagh, tadukan*
China	*Cina*, DGWG, *pa chiam, tsai yuan chan*

Source: Maheshwari (1997).

Land Management for Agriculture in the Koraput Region

Agriculture is mainly a family affair and happens in two kinds of lands, viz. wetland and dryland (*fig*. 88). Dryland could be a mountain slope or a field in the plains where water availability is low and is mostly rainfed. Wetlands are fed by streams. Rice is the only wetland crop cultivated by all four communities. On the mountain slopes, millets, dryland rice varieties, sorghum or maize are cultivated. The drylands in plains are mostly used for vegetable cultivation.

TERRACE CULTIVATION

Streams are manipulated to create wetlands for rice cultivation (*fig*. 77). Terraces are constructed across the flowing streams with boulders as barriers thus levelling small plots of land for rice cultivation and also for managing the flow of water. One dose of farmyard manure is applied as basal manure. It is commonly observed that the stream widens in areas of cultivation where its existence is barely noticeable and narrows to a normal stream immediately after the cultivation area. The significance of this system is:

i. The nature of the stream is unchanged. The water is not dammed or stored and hence much of the water utilised for agriculture is returned to the stream itself. The water available for the crop can still be controlled by constructing temporary bunds at the side of the field that receives water from the upper terrace plot. The diverted water moves by the side of the field on to the plot downstream.

ii. Since no synthetic fertiliser is used, the water quality is not much altered.

iii. This system eliminates the need for applying external energy for pumping out water from the water source.

SHIFTING CULTIVATION

The Koraput region is known for its shifting cultivation, generally referred to as *podu*. The mountain slopes are cleared off vegetation except a few cultural keystone tree species such as *Mangifera indica, Artocarpus heterophyllus* and *Caryota urens* (Franco et al. 2008). After this, the ground level vegetation and left-offs are burnt and the ash is left behind to enrich the soil (*figs.* 78 & 79). The tradition was to cultivate the land for three consecutive years and leave it fallow for another 20 years for the forest to regenerate. At present, increase in population has made this practice impossible except in the Bonda hills. The land is cultivated consecutively for three years and then left fallow only for two-three years or cultivated continuously. However, the condition is slightly different with the Bonda community where the *podu* plots adjacent to the village are generally cultivated for three years and then left fallow for another three years while the forest lands far away from the village are cleared and cultivated for three consecutive years and then left fallow for 10-15 years. The *podu* fields in the regions inhabited by the Kondh, Poraja and Gadaba permanently belong to the family that initially cleared the area while in the Bonda community; the ownership enjoyed by the family that cleared the forest patch ceases to exist four-five years after the area has been left fallow. This period is determined by the pace of regeneration of vegetation and the disappearance of the characteristic features of a field. Hence the period of four-five years is not a finite one and largely depends on the pace of forest regeneration as the Bonda is emotionally attached to the field as long as the effort taken by his family initially to clear the forest lingers in the plot.

Measurements

The indigenous measurements used by the communities to measure agricultural produce are *oda* (1 kg), *mano* (3 kg), *dala* (20 kg), *putti* (60 kg) and *kavadi* (40 kg).

Crops cultivated

Dhano (*Oryza sativa*), *mandia* (*Eleusine coracana*) and *suan* (*Panicum miliaceum*) are the three major crops cultivated by the communities. *Kangu* (*Panicum miliare*) which was a major crop is rarely cultivated these days. Apart from these, maize, sorghum and vegetables such as brinjal, chilli, cabbage, cauliflower, tomato, groundnut, beans, pumpkin, ginger, turmeric, red gram, black gram, Bengal gram, cucumber and bitter gourd are also cultivated. *Olsi* (*Guizotia abyssinica*) is a major cash crop cultivated in low lying *podu* fields.

The major crop profile of communities studied is provided in Tables 27-30 (*figs*. 80-87). Note that the remark "extinct" is used to denote locally extinct species within the community, and given the practice of informal seed exchange between communities, they could exist in other communities.

Table 27: Major Crops Cultivated by Kondh

Name of the Crop	Name of the Cultivar	Crop Duration (Months)	Yield (kg/kg of Seeds)	Land	Special Attributes/ Remarks
Dano	*Balu kuli*	8	18	Wetland/ semi-wetland	Scented
	Batta lochoi	5	30	Mountain slopes/rainfed	Extinct
	Heeya pidika	8	18	Wetland/ semi-wetland	Scented, retains smell even after two years of harvest. Swells more on cooking
	Kaja kuli	8	18	Wetland/ semi-wetland	Scented rice
	Lochoi	8	18	Wetland	Sweet; grains can be stored for a year. Good for porridge
	Maha Kuli (Chipti)	5	18	Wetland	Less breakage

	Oona lochoi (san lochoi)	8	18	Wetland/ semi-wetland	For porridge
	Saphur lochoi	6	40	Wetland	Sweet scented
Mandia	*Badu mandia*	8	70	Mountain slopes/rainfed	Porridge will be watery
	Beeri mudha	5	20	Mountain slopes/rainfed	-
	Dosra Bodi	5	60	Mountain slopes/rainfed	-
	Kaja mandia	5	20	Mountain slopes/rainfed	Tall plant with long inflorescence
	Mandia koori	6	18	Mountain slopes/rainfed	Early harvest
	Todo mandia	5	20	Mountain slopes/rainfed	Forms cakes when cooked
Sua	*Kaja kahoda (bodo sua)*	8	30	Mountain slopes/rainfed	-
	Kurji kahoda (mami sua)	3	15	Mountain slopes/rainfed	Sweet taste
	Ginjeri kahoda	3	15	Mountain slopes/rainfed	-
	Kodru kahoda	8	30	Mountain slopes/rainfed	-

Table 28: Major Crops Cultivated by Poraja

Name of the Crop	Name of the Cultivar	Crop Duration (Months)	Yield (kg/kg of Seeds)	Land	Special Attributes/ Remarks
Dano	*Badham lochoi*	6	-	Wetland	Extinct
	Balu dhan	7	20	Wetland	Black paddy, resembles *nini dhan*
	Bata dhan	4	12	Mountain slopes	Hard and coarse grains, takes time for cooking

	Bodi dhan	8	5	Wetland	White, small grains
	Chipti	5	18	-	Less breakage
	Kalia septi	4	12	Semi-wetland	Black grains, white rice
	Kandul gatti	8	20	Semi-wetland	Extinct
	Karandi	7	20	Semi-wetland	Large grains, require more time for cooking
	Lochoi	8	18	Wetland	Less breakage
	Mauli dhan	7	20	Wetland	-
	Nini dhan	8	20	Wetland	Small grains, less breakage
	Padha dhan	6	12	Rainfed	Extinct
	Pathor losa	7	20	Wetland	-
	Saphur dhan	8	20	Wetland	Sweet, less breakage
	Umria sudi	8	18	Wetland	Extinct
Mandia	*Bodo mandia*	6	60	Mountain slopes	1 kg of flour makes gruel for 10-12 people
	Badu mandia	8	36	Rainfed	-
	Budel mandia	8	30	Rainfed	-
	Chilli mandia	5	30	Mountain slopes	1 kg of flour makes gruel for 20-25 people
	Dosra bodi/ Dosra budeli	4	80	Rainfed	-
	Jona mandia (mami mandia)	3	80	Mountain slopes	-
	Kalagotti	5	240	Mountain slopes	Plant, inflorescence and seeds black, nodes alone white
	Machdim mandia	5	120	Rainfed	-
	Mardi mandia	7	80	Mountain slopes	Not sour
	Martha	8	30	Rainfed	Tasty
	San mandia	5	30	Rainfed	-
	Sona deyi	5	240	Mountain slopes	Smaller grains but higher flour yield

Sua	*Bod sua*	9	1,200	Mountain slopes/ rainfed	High yield in mountain slopes, less yield in rainfed land
	Kalia sua	5	300	Rainfed	-
	Kosum thappa	4	60	Mountain slopes	For periods of food scarcity
	Mami sua	4	20	Mountain slopes	Used during periods of food scarcity
	Sacra sua (masli sua)	7	20	Mountain slopes	For periods of food scarcity, black in colour

Table 29: Major Crops Cultivated by Gadaba

Name of the Crop	Name of the Cultivar	Crop Duration (Months)	Yield (kg/kg of Seeds)	Land	Special Attributes/ Remarks
Dano	*Balu varsil*	4/5	-	-	Extinct
	Batta dhan	4/5	-	-	Extinct
	Bodi varsil	4/5	-	-	Extinct
	Buduma	8	15	Wetland	Extinct
	Holdi saphur	8	15	Wetland	Extinct
	Kandul katti	8	15	Wetland	-
	Karandi	5	80	Wetland	-
	Kibadhan	6	2	Mountain slopes	Tasty
	Lochoi	5	18	Wetland	-
	Matti mami	6	15	Mountain slopes	-
	Mauli varsil	-	-	-	Extinct
	Nini varsil	-	-	-	Extinct
	Olsi saphur (gundri dhan)	4/5	15	Wetland	Scented, extinct
	Podha buduma	6	15	Mountain slopes	-
	Saphur	4/5	18	Wetland	-
	Siptil (sipti/chipti)	4/5	20	Wetland	Smaller rice grains
	Sudi varsil		-	-	Extinct

	Veend varsil	4/5	20	Rainfed	-
	Roktha kosa	4/5	20	Wetland	Red rice, tasty
Mandia	*Bodel mandab*	5	12	Rainfed land	-
	Bodi mandab	6	120	Rainfed	-
	Bodo mandab	6	100	Rainfed	-
	Silli mandab	6	180	Rainfed	-
Sua	*Baer sua (Bod sua)*	6	40	Rainfed	-
	Kora sua	4	10	Rainfed	-
	Masli sua	6	40	Mountain slopes	-
	Pudhey sua (mami sua)	3	25	Mountain slopes	-
Kangu	*Kangu*	3-4	20	Mountain slopes	-

Table 30: Major Crops Cultivated by Bonda

Name of the Crop	Name of the Cultivar	Crop Duration (Months)	Yield (kg/kg of Seeds)	Land	Special Attributes/ Remarks
Dano	*Banda saphur*	6	50	Wetland	Awns up to 1 inch long, scented, coarse grains
	Biri kaer	6	30-40	Mountain slopes	-
	Bobai kaer	6	60	Rainfed	Black grains
	Lasaer kaer	6	80	Wetland	-
	Lokmo kaer	3	50-60	Rainfed	Grains black at one end
	Musri kaer	3	60	Rainfed	Very small grains
	Nealli	3	40	Wetland	-
	Saphur	3	50-60	Rainfed land	-
	Seethai	6	80	Wetland	Round grains
	Piri amo	6	50	Wetland	-

Mandia	*Dosra samey*	6	50	Mountain slopes	-
	Kalia gunti	6	20	Rainfed	-
	Puthru samey	6	50	Mountain slopes	-
Suan	*Budhuki*	6	150	Mountain slopes	-
	Gusoguri	6	120	Mountain slopes	-
	Le dokri	6	250	Mountain slopes	Black grains
	Maalem	6	200	Mountain slopes	-
	Mogondri	6	250	Mountain slopes	-
	Purga lamri	6	120	Mountain slopes	-

Traditional Knowledge Associated

SEED SELECTION, CONSERVATION AND THE ROLE OF WOMEN

In all the four communities studied, the seeds are selected carefully from stands that have performed well in terms of yield and quality. The selected germplasm is then hand-threshed (in the case of rice and millets) and stored carefully. Seeds meant for sowing are stored in pots, bamboo baskets or gunny bags. Selected pumpkin and bottle gourd fruits meant for seeds are stored on rooftops, exposed to rain and sunlight. In villages where the traditional way of conserving seeds persists, it is the woman who is the caretaker of the seeds. In the Kondh community, the eldest woman of the family is supposed to take care of the seeds in a special earthen pot. Any damage to this seed pot and its contents is considered as a bad omen for the family concerned. This practice is still followed in the village of Chikalmari while in Putisil it is still maintained by the women but in gunny bags or plastic sacks.

In the Gadaba and Poraja communities too the women take care of the seeds, but there is no belief attached. In the Gadaba community, the housewife is supposed to hand over the seeds maintained by her to her man so that he initiates sowing. While doing so, it is observed that the women carry the seeds in the free end of the saris worn. In the case of the Poraja and Gadaba, Grains for consumption are stored in huge bamboo baskets plastered with mud and cow-dung externally that go up to 6 × 3 × 3 ft in dimension. In the Bonda, any woman of the family can safeguard the seeds in *thumba* (bottle gourd), *thudi* (bamboo baskets) or *handi* (earthen pot). Since Bonda families are generally nuclear in nature it is the housewife who is responsible for maintaining the seeds. All these underline the important role played by women in agro-biodiversity conservation, which has been acknowledged worldwide (Trager 1982; Akhter et al. 2010). In the Koraput region, these women who are the caretakers of seeds are now known as "*Bihani Maa*" meaning "seed mother", a term coined by the NGOs operating in the area (Mohanty 2009).

PEST MANAGEMENT

The inherent resistance offered by the traditional cultivars to insect pests and diseases renders the crops nearly maintenance free. Though there is no synthetic fertiliser or pesticide applied, the diversity of crops cultivated ensures that there is no major outbreak of pests and loss of crop due to pests is nearly zero. All the communities in the Koraput region are known to rear colony spider (*Stegodyphus sarasinorum* Karsch) in their fields. The colonies are picked from elsewhere and translocated to the agricultural fields. This practice helps in controlling insect pests. Moreover, colony spiders and other spiders have also been observed on the structures left behind in the fields after rituals. The communities perform rituals in the name of avoiding evil glances at the crop. In these rituals, branches of trees are cut and erected in the fields. The Kondh erect ladder-like structures in addition to the branches. Such structures are left behind in the field and serve as bird perches and also harbour colony spiders (*figs.* 102 & 103).

WEATHER PREDICTION

The Kondh farmers have a unique way of predicting rains. If the *laden* (long black earthworms) are seen to move from low-lying areas to highlands, rain can be expected and vice versa. This observation of earthworm movement as a weather indicator seems to be quite effective.

Rituals, Festivals and Other Beliefs Concerned with Agriculture

The main festivals/rituals associated with the agriculture are provided in Table 31.

KONDH

Mag Porv/Bicha Porv

The festival of *Mag porv* or *Bicha porv* is celebrated in the month of Magh (February), prior to the commencement of sowing of seeds (*figs.* 89-91). It is a two-day festival.

Table 31: Festivals/Rituals Associated with Agriculture

Festival/ritual	Community	Month	Purpose
Choith porv	All communities	March-April	Spring festival
Annual Ritual Hunt	All communities	March-April	Hunting
Mag porv	Kondh	February	Initiation of sowing
Bali porv	Kondh	April	Initiation of sowing
Nondi porv	Poraja	February	Initiation of finger millet cultivation
Kandul baja	Poraja	November-December	Consumption of black gram and red gram
Daal gada	Gadaba	August	Ward off pests
Kida viru	Gadaba	September	Ward off pests
Bion jatra	Gadaba	March-April	Initiation of sowing
Jona nua	Gadaba	July	Assuring good harvest of maize
Osa porv	Bonda	September	Harvest of maize
Kumbogda	Bonda	October	Consumption of rice and *sua*

On the first day, the Dishari along with the Jani and the other villagers go to one end of the village and sow some seeds, mostly rice in three pits (*Gumum pooja*). *Londha* (brew) is poured on these pits. After this, an elderly woman from the village applies a *tikka* (tilak) on the forehead of the Dishari and the Jani. The duo then proceed to the Jakeyr of the Tada Pennu (Earth God) who is represented by a small cave-like structure of around one square foot in the sacred grove, with a small stone slab as door. All the while, the Jani will be carrying a bamboo stick with two bamboo baskets tied to one end. One of these baskets will be containing sample rice seeds from the lot meant for sowing and the other basket would contain an earthen lamp burning in castor oil. To the other end of the bamboo stick, an earthen pot containing cooked rice is tied. Once the duo reach the Jakeyr, the stone door of the cave-like structure is opened and the earthen lamp is transferred inside. The lamp is kept burning till the end of the festival.

In the Jakeyr, rice is cooked in a mud pot as well as in a bamboo stalk of around 1 ft length. Two types of drums, viz. *dapu* (earthen pot tied with cow's skin) and *liseni* (goat hide tied around a wooden frame) are beaten. The night is designated as a free night where any individual from the village is permitted to stealthily take away the *kandul* (red gram) from any individual's field and fry it in the Jakeyr. These people then have a night halt at the Jakeyr. Next morning, people gather at the Jakeyr, and the Dishari, along with the Jani, begins the ritual where elaborate chants are involved. He then consumes liquor, sprinkles some seeds on the floor and offers liquor to the deity. Following this, the Bejeyni of the village begin their rituals that consist of chants that continue for a minimum of a couple of hours. All along the chant, they keep consuming liquor, sprinkling seeds on the floor and offering liquor to the deity. Following this, a chicken is sacrificed. Every year, a chicken is sacrificed; once in three years a pig is sacrificed and once in 12 years, a cow is sacrificed. The blood of the sacrificed animal is offered to the god and then mixed with the seeds. The seeds are then sprinkled in the air and the villagers collect them by spreading clothes.

The collected seeds are taken back home, mixed with other seeds and sown in the field. In the event of the field being not ready for cultivation, the seeds are dumped in the compost pit, since the compost too ends up in the field. The festival ends with a community feast at the village where rice and the meat of the sacrificed animal is served, followed by the serving of *londha* to the villagers by the Dishari and a community dance.

Choith Porv

This festival is celebrated in the month of Hirey (March-April) and hence is also called as *Hirey porv*. Traditionally, this festival used to happen for one full month in many stages. The first part of the festival is concerned with bird hunt and the second with animal hunt. Both these ritualistic hunts are collectively called as Annual Ritual Hunt (ARH). Animals, mostly herbivores are hunted during the ritual. The festival is common to all tribal communities in the Koraput region and is celebrated in a similar fashion but for minor cultural variations. The significance of ARH is discussed under a separate heading.

Bali Porv

Bali porv initiates the second crop season. The festival begins on a Monday following *Choith porv* and continues for nine days. A shed made of leaves from *Caryota urens* and decorated with flowers of *Plumeria rubra* is erected at the centre of the village. Sample seeds from the seed lot meant for sowing are sown in small earthen pots and placed inside this shed. These seeds are watered every evening, accompanied by rituals by the Bejeyni (*figs*. 92-94). Dance is performed by the Bejeyni every evening except on the Wednesday. The dance is quite a spectacular one with the Bejeyni dressed up in white sarees with flowers of *Plumeria rubra* adorning their foreheads as a wreath. Flowers of *Plumeria rubra* are preferred since they are considered as symbols of purity. An indigenous snack called as *pitta* (steam-baked cake) is made out of finger millet flour on the final day of the rituals and consumed.

On the first Saturday of the festival, there is a ritual where the village deity (an invisible force) is symbolically requested to leave the village by offering a chicken and coconut as sacrifice by the side of the path leading to the village. On the last day of the festival, the seedlings are brought out of the shed and worshipped with flowers of *Plumeria rubra* and by sacrificing a chicken. The pots containing seedlings are then taken to the pond or stream and washed away. One flower of *P. rubra* is given to each family of the village as a token of good omen that would invoke a good crop. The festival ends with a water game when the villagers throw water on others, mostly on the members of the opposite sex. The festival can be considered as a method of testing the viability of the seeds, since it is understood that traditionally the failure of seeds to germinate was considered as a bad omen and the seed lot represented by the poorly germinated seeds will be discarded. This festival is equivalent to the *Mulaipari* festival celebrated in northern Tamil Nadu (personal observation).

PORAJA

Nondi Porv

Nondi porv happens in the month of Magh (February) and is celebrated in honour of the Nondi God and Goddess, who occur as a divine pair (*figs*. 97 & 98). Though these gods are considered formless, the Poraja attribute a crude rat-like form to them during this festival. The idols are made of mud collected from termite hill, existing in or near the Nondi Jakeyr. The ears of the idols are represented by seeds of cucumber. The gods are adorned with gold-coloured ornaments bought in local *haat*. These gods are kept in a square-shaped *nondi chengatta* (bamboo basket) which is bordered by four stalks, one in each corner. These stalks are interconnected with a thread from which flowers of *semi* (*Lablab purpureus*) are hung. A lamp burning in castor oil on an earthen pot is also placed in the basket.

On the day of the festival, the basket is kept in heap of *dehra* (finger millet) at the house of a Poraja belonging to the Padi clan. An indigenous drawing, depicting the various natural resources such as land, stream, birds, cattle and crops is drawn on the wall existing behind the place where the basket is kept. Above the basket, a hoop is hung from the ceiling of the house. Finger millet spikes are found hanging from the sides of the hoop. The entrance of the house is also decorated with strands of finger millet. A black chicken is sacrificed before the gods and its head is dropped in the basket. The basket is carried from house to house and finally drowned in a stream in the evening. It is believed that the Nondi god is responsible for assuring a good harvest of finger millet. The festival marks the beginning of finger millet cultivation.

Kandul Baja

This festival is usually celebrated in the month of Pond (November-January) at the Pand Jakeyr, presided by the deity *Pand depta*. The festival initiates the consumption of *miri* (black gram) and *kandul* (red gram). On the day of the festival, each family of the village contributes a little of *kandul* and *miri* which is collected by volunteers and taken to the Jakeyr. At the Jakeyr, both these grains are roasted separately and offered to the deity along with rice. It is believed that the failure to host this festival would bring down evil on the village.

Choith Porv

During the festival of *Choith porv*, a goat, chicken and a coconut are sacrificed to the deity, mostly Gangamma and rituals are done in honour of the deity. The Annual Ritual Hunt (discussed separately) commences on the next day when the male members of the village go on a bird hunt called *bod bet*. The birds downed are tied to bamboo canes decorated with flowers of *tingri* (*Firmiana colorata*) and hence this hunt is also called as *tingri bet*. In the evening, the male members gather in the sacred grove for a ritual honouring their weapons (*figs*. 99 & 100). A small patch, around 1 sq ft is cleared and plastered with cow-dung on which four small squares are drawn. Rice is placed on each of the squares and rituals are offered with an egg, twigs of castor and twigs of *F. colorata.* The egg is then broken and smeared on the weapons. This is followed by dance and in the morning, the group returns again to the same place and an egg is placed on the centre of the four squares and the male members try to shoot down the egg from a distance of 7 arrows from the egg. This ritual initiates the game hunt.

Towards the end of the festival when the hunt is over, a sub-festival called *Kotna kavoni* occurs. The whole village goes into the forest in search of tubers, mostly *Dioscorea spp.* These tubers are brought back and cooked by the women-folk while the men initiate ploughing in the field. The women carry the cooked food to the men and a feast is held in the field. In the afternoon, the male members go on a mini-hunt that marks the end of Annual Ritual Hunt after which people gather at the Jakeyr where the Dishari conducts the rituals and sprinkles the seeds in the air. The seeds are collected and sown by the people on the next day.

GADABA

Daal Gada

This festival happens in the month of Bondapan (August). Branches of *Syzygium cumini, Semecarpus anacardium* and plants of *Sida acuta* are cut and erected in the agriculture field. Rituals are done by the Dishari. It is believed that the rituals keep the insect pests away from the field.

Kida Veeru

This is another festival conducted for preventing insect pests in the field in the month of Osa (September). Mud casts of insect pests are done and rituals done by the Dishari by the side of the path leading to the village. A goat is sacrificed along with a coconut.

Bion Jatra

This festival happens in the month of Choith (March-April). *Bion jatra* is a part of the larger festival called *Choith porv.* The villagers gather under the aegis of the Dishari at the sacred grove of Gangamma with *pendham* and liquor (branded liquor in recent times). Liquor is offered to Gangamma and also sprinkled on the seeds and these seeds are handed over to the villagers for sowing. The date for the commencement of the Annual Ritual Hunt is also announced on this day. This festival is a Gadaba equivalent of the *Bicha porv* of Kondh.

Jonanua

This festival happens in the month of July. Tender leaves of maize plants and *sua* are cooked with rice, tamarind and pumpkin and offered to the gods along with a chicken sacrifice in the home garden. A little of the prepared food is kept on a *Bauhinia vahlii* leaf and offered to the forefathers by the side of the path leading to the village. This festival concludes before the maize crops are ready for harvest and is believed to assure a good yield.

BONDA

Choitharkey

Choitharkey is equivalent to the *Choith porv* celebrated by other communities. On the first day of the *porv*, *Pothor pooja* is done as a mark of respect to the *sadhur*. Though the meaning of this ritual is not known to the present generation of Bonda, the presence of long stone slabs that represent their ancestral spirits and the manner in which the rituals are performed suggest that this ritual is done to show respect to the ancestral spirits. During this ritual, a chicken, pig and egg are sacrificed. On the next day which would be a Sunday, Mango collected from the wild is brought and rituals are done. On the Monday and Tuesday, the Bonda go on hunting for game and on Wednesday, there is a break effected by a ritual called *hirang* where the male members going on hunting shoot arrows on a *simili* (*Bombax ceiba*) tree. It is believed that this improves their skill of archery and also brings good luck to the archer. The hunt resumes again on the next day and would continue till the end of the festival. Towards the end of the festival, there is another sub-festival called *Bion porv* when the villagers bring sample seeds from the seeds meant for sowing in small earthen pots or bottle gourds and hang them in the tree on the *sadhur* of their respective villages. These pots and gourds continue to remain in the tree for one to two months even after the ending of the festival. *Choith porv* continues for one full month during which no stranger is allowed to enter the village and no villager is permitted to stay away from the village during nights. Use of oil, both on the body as well as for cooking is prohibited. Similarly, the Bonda custom prohibits cooking of food inside the home during this festival.

Osa Porv and Kumbogda

Apart from the *Choith porv*, two other festivals, viz. *Osa porv* and *Kumbogda* are celebrated by the Bonda. *Osa porv* is celebrated during the month of *Osa* (September) and initiates the harvest and consumption of maize. *Kumbogda* is celebrated in the month of *Dosra* (October) and initiates the consumption of newly harvested rice and *suan* (*Panicum miliaceum*).

Bonda beliefs associated with agriculture

i. While on the way to the field, it is customary to abstain from speaking with any people whom the Bonda would meet.

ii. While working on the field too, the Bonda customs prohibit people from speaking to each other. This stands in contrast to the customs practised by the other three communities where songs accompany agricultural activities. However, this practice also highlights the high reverence accorded to agriculture by the Bonda, since the Bonda are also known to maintain silence during other rituals and invocations.

Significance of Annual Ritual Hunt

Hunting is known to be associated with the human evolution. Even Neanderthals who lived during the middle-palaeolithic period had specialised in hunting (Adler et al. 2006). Though hunting does not form the primary food source for most communities today, it is still found to be associated with indigenous societies. In the Koraput region, hunting is still associated with the indigenous culture. Here, hunting does not form the primary food source anymore. Yet, it is an essential ritual of the festival *Choith porv*. Since this hunting happens as a ritual annually, it is called as Annual Ritual Hunt (ARH). During ARH mostly herbivores are hunted. Animals such as bears, leopards and snakes are not hunted. Leopard are not spotted these days due to poaching and the bear population is noticeable due to the taboo on its killing that was prevalent once. However, my observation shows that in the recent years, communities in the Koraput region have started hunting bears too, possibly because of the nuisance caused to the crops.

In the context of the past where the agricultural fields of the communities were closer to large patches of forests that sustained a diverse range of herbivores, the ARH was effective in thinning the herbivore population that would otherwise invade the agricultural fields. Hence, ARH has played a major role in regulating and managing herbivorous wildlife otherwise the population of which can grow rapidly and cause crop loss to the community, resulting in food insecurity. However, in the present scenario, ARH seems to be destructive with all the villages reporting a decline in the availability of game. This could be understood from Table 32 that compares the game hunted before the year 1996 and in 2005.

Sambar deer, barking deer and spotted deer were neither hunted nor spotted during ARH in 2005 and after that. This shows that the availability of larger game animals has depreciated. Hence this practice today is simply a cultural relict of what was once a practice of managing wildlife that affects agriculture.

Table 32: Comparison of Animals Hunted before 1996 and in 2005

Animals Hunted before 1996	Spotted deer, barking deer, sambar deer, hares, mongoose, peafowls, wild boar, jackals, porcupine, jungle fowl
Animals Hunted in 2005	Hares, mongoose, peafowl, wild boar, porcupine, jackals

Present scenario

The Odisha state government policy promotes organic farming in the tribal villages (Government of Odisha 2008). It also speaks about conservation of indigenous seeds. However our observation shows that hybrid seeds of high-yielding nature are provided through the *panchayat* block administration. Synthetic fertilisers are also made available for a highly subsidised rate (free in the case of Bonda). Though these improvised seeds and synthetic fertilisers have increased production in the villages where they were implemented, the input costs have risen along with disease and so is the pest incidence, as reported by the communities. As a result, the net profit remains the same or marginally higher. The state agricultural policy speaks about exploring market facilities for the organic agricultural products produced by the tribal communities. But the policy is silent about the status of indigenous seeds used by the tribal farmers. The availability of free or subsidised rice through the Public Distribution Service (PDS) has decreased the dependence of the communities on agriculture for food and has also changed the food preferences of the people. As a result, "white rice" or polished rice, smaller in size is preferred statewide, resulting in lesser demand for the coarse grained rice produced by the tribal communities. For the tribal farm produce to be successfully marketed, the state government should formulate appropriate policies and strategies.

8

Indigenous Technologies

> The experimental process that originally led to the manipulation and combination of the morphologically dissimilar plants, and the discovery of their unique chemical properties, is far more than the phrase "trial and error" suggests. — Davis 1995

INDIGENOUS technologies are those that had evolved *sui generis* in Indigenous communities. These technologies enabled the communities to utilise the natural resources using simple, but multi-step processes. Though simple, these technologies often serve as means to cope up with extreme climatic conditions and bear testimony to the scientific knowledge of the communities (Roy et al. 2004).

Fermentation of food products is carried out so as to obtain solid or liquid fermented food products by employing traditionally selected microbial organisms and a wide range of plant/animal products (Adebesin et al. 2001; Henkel 2005). Sometimes, the fermented products are distilled to obtain liquors with stronger alcohol content. Fermentation maximises the protein content of the food. Mildly alcoholic beverages also play an important role in indigenous culture by sustaining community bonds. Beverages and distilled liquor are of immense importance in traditional rituals, festivals and other important occasions. In India, most of the traditional food recipes employ the technique of fermentation. Fermented drinks obtained from *Caryota urens, Borassus flabellifer, Cocos nucifera, Oryza sativa,* etc. are popular among the indigenous as well as the mainstream communities. Platt (1955) reports five main groups of alcoholic beverages from Africa, viz. fermented honey water, fermented fruits and juices, fermented sap from various species of palm and a bamboo, drinks made from milk, and the beers.

The fermented drinks might appear as an element of luxury when viewed superficially. But a careful observation shows the nutritional and cultural value of them (Platt 1955; Bahiru et al. 2001). Besides, the beverages are also sources of income to the indigenous people (Abegaz et al. 2002). Backstrand et al. (2001) show that the moderate consumption of *pulque*, a central Mexican beverage can provide many vital micronutrients to the otherwise nutrient deficient pregnant women. Other studies too confirm that traditional fermented beverages are healthy food due to their mild alcoholic content (Oladeinde et al. 2002). Plants that are sources of beverages are also known to be accorded a high status in indigenous communities (Franco et al. 2008).

This chapter deals with the indigenous technologies on food and oil extraction, documented from the four communities.

Beverages

All the four communities prepare two kinds of brew, viz. the *pendham* and *londha*. *Pendham* can be prepared from both rice and finger millet while *londha* is prepared from finger millet alone. In the preparation of *pendham*, rice is coarsely pound and cooked. It is then allowed to cool. After cooling, the cooked rice is put in an earthen pot and added with water. To this, barks from *Plumbago zeylanica, Alstonia venenata* and *Holarrhena pubescens* are crushed and added. Sometimes, these barks are crushed, added to rice flour and made into balls/tablets and stored (*fig*. 104). The mixture is then covered with leaves of *Bauhinia vahlii* and allowed to stand in a corner of the house for three to four days. Fermentation takes place during this period and the brew is ready for consumption after being filtered. The process is almost the same for the preparation of *pendham* with finger millet too, except that finger millet flour is mixed with a little of water and left overnight before being added with more water and plant barks. However, there are cultural variations in the preparation of these two beverages in other parts of Odisha (Patnaik 1976).

For preparing *londha*, clean finger millets are collected in a bamboo tray and wetted with water. It is allowed to stay as such for two days when the finger millets would germinate. The plumules are now separated from the finger millet seeds and seeds alone are subjected to fermentation after adding with the barks of the three plants mentioned above. *Pendham* and *londha* form an important part of tribal culture as no festival or occasion will be complete without it.

Distilled Liquor

Sagur is the generic name applied to all distilled beverages made from a variety of sources (*fig*. 65). Bonda are capable of distilling *sagur* out of any food material. However, they generally distil it out of tamarind fruits, *mahua*, jackfruits, mango, *salop* and from the pseudocarps of *Anacardium occidentale.* The fruits are collected and soaked in water for a week or until it gets fermented. The whole locality of Bonda hills smells like liquor during the mango and jackfruit season. The fermented fruits are then collected and boiled in an earthen pot (pot 1), by the side of a stream. There is another smaller earthen pot (pot 2) placed in inverted position above this, bearing holes on its bottom, which is plugged with clay or cow-dung. There is an opening at one side of the upper pot that leads through a bamboo pipe into a third earthen pot placed in running water. When the pot one is heated, vapour from the fermented fruits move to the second pot and then to the third pot where they condense owing to lower temperature, brought about by the running water. This constitutes the *sagur* and is consumed as such or re-distilled for higher intoxicating potency. In the same way, *sagur* can also be prepared from the toddy obtained from the *solap* tree (*Caryota urens*) as well as from *pendham* and *londha*.

Oil Extraction

Oil is mostly extracted from *Pongamia pinnata, Casearia graveolens* and *Ricinus communis*. Of these, the technology followed to extract oil from castor differs from the former two. *Pongamia* or *Casearia* seeds are dried under the sun and then coarse ground using the traditional pounder. The coarse flour is then steam cooked in an earthen pot with a porous bottom, placed on another earthen pot containing water. Later, the steam cooked flour is again sun-dried and tied in pouches made of *Bauhinia vahlii* leaves or of late, porous plastic bags (*figs*. 105-111). There is a permanent socket carved in a common tree of the village at about 2 ft from the ground. A wooden plank with furrows is placed beneath the socket and the bag/ pouch containing the flour is placed above it. A wooden log, about 8 ft in length is inserted into the socket at its unforked end and the villagers apply pressure on the forked end of the log manually. Oil squeezes out from the ground seeds under the pressure and flows through the furrows on the wooden plank onto a vessel placed so as to collect the oil. The matter left after oil extraction is used as manure in the agricultural fields.

In the case of castor, the fruits are boiled and then dried. Seeds are now removed and coarse ground using traditional pounder. The flour is now boiled with water in an earthen pot. Oil separates from the flour and floats at the top of the solution which is collected using hen's feathers. The process of oil extraction is labour-intensive and is fast eroding due to easy availability of branded oils in the nearby markets. However, oil extraction has got a few benefits such as control over the quality, manure that is obtained as by-product and moreover it does not involve any cash expense.

Fig. 1: A podu field bordered by a patch of secondary forests in Bonda hills.

Fig. 2: Community Protected Forests are a common feature in Koraput district. In contrast, hills cleared of vegetation for podu is equally common in the region.

Fig. 3: Olsi is an important cash crop cultivated for its oil seeds. Its cultivation enhances the aesthetic appeal of the landscape.

Fig. 4: The vegetation of the cremation/burial grounds represents both conservational and cultural values of the community.

Fig. 5: The Kondh villages are constructed in a linear pattern, always in the east-west direction.

Fig. 6: Poraja and Gadaba villages are found either at the foothills or in the valleys. A Community Protected Forest is always seen in the vicinity

Fig. 7: Poraja and Gadaba villages are found either at the foothills or in the valleys. A Community Protected Forest is always seen in the vicinity

Fig. 8: Bonda villages are nested in the mountains, surrounded by secondary forests.

Fig. 9: Making a yoke requires great skill but simple implements.
It is vital to carry loads for greater distances in the absence of motorized or animal driven vehicles.

Fig. 10: Poraja women are quite forthcoming and are enthusiastic participants in any conversation.
Their charm is only enhanced by traditional ear and nose rings.

Fig. 11: Wedding is a simple affair in Poraja villages. A new dress and a scarf on the head is all that differentiates the bride from her friends. The photograph shows a Poraja bride and her pal standing by the side of a sadhur.

Fig. 12: Bonda women are known for their elaborate ornaments. The more the number of osangoo (neck rings) that a woman wears, higher is her status and pride. Their shaven heads are decorated with colourful beads. The younger generation adopts a blend of modern attire and traditional ornaments.

Fig. 13: Bonda women are known for their elaborate ornaments. The more the number of osangoo (neck rings) that a woman wears, higher is her status and pride. Their shaven heads are decorated with colourful beads. The younger generation adopts a blend of modern attire and traditional ornaments.

Fig. 14: Bonda huts are simple and made from locally available materials. A pig sty or a goat pen is seen attached or nearby.

Fig. 15: The haat or the weekly market is an integrated part of tribal culture. Local goods are sold or bartered for goods from outside.

Fig. 16: Both the young and elderly couples eagerly look forward for the haat. Gifts and ornaments from the haat play an important role in their love life.

Fig. 17: *Alstonia venenata* — used for suicide as well as for medicine.

Fig. 18: *Ardisia solanacea* — a source of edible fruits and medicine.

Fig. 19: *Argyreia nervosa* — cures indigestion and reduces swellings.

Fig. 20: *Bauhinia vahlii* — an important NTFP as well as yields edible seeds.

Fig. 21: *Bauhinia variegata* — flowers are used for culinary purposes.

Fig. 22: *Bombax ceiba* — a keystone cultural species; blooming of which marks the arrival of spring.

Fig. 23: *Breynia retusa* — a major source of medicine to treat dysentery.

Fig. 24: *Buchanania lanzan* — the kernels form a NTFP; pericarp is eaten by the communities.

Fig. 25: *Caryota urens* — a cultural keystone species; toddy extracted from its inflorescence is a valuable social drink as well as a food for lean season.

Fig. 26: *Casearia graveolens* — The Bonda use oil from the seeds for culinary purposes; others do not.

Fig. 27: *Cassia fistula* — a multipurpose tree used for purposes ranging from medicine to rituals.

Fig. 28: *Cinnamomum caudatum* — its aromatic bark is used as febrifuge.

Fig. 29: *Clematis roylei* — leaves are used as a decongestant.

Fig. 30. *Costus speciosus* — the rhizome is used for cough, stomachache and migraine.

Fig. 31: *Cryptolepis buchananii* — latex yielding plant is used mainly as a galactagogue.

fig. 32: *Dalbergia ougeinensis* — timber is used for house construction.

Fig. 33: *Ficus semicordata* — fruits relished upon by all communities.

Fig. 34: *Firmiana colorata* — the colourful flowers are an integral part of the festival of choith porv.

Fig. 35: *Furcraea foetida* — an exotic plant used as medicine.

Fig. 36: *Hydrocotyle rotundifolia* — whole plants when chewed, can cure white spots on the tongue.

Fig. 37: *Indigofera cassioides* — a plant of medicinal and edible value.

Fig. 38: *Mallotus philippensis* — fruits are sold as a NTFP and also used as a source of natural dye.

Fig. 39: *Merremia vitifolia* — leaves are used to cure stomach ache.

Fig. 40: *Mimosa pudica* — the touch sensitive leaves are believed to induce sleep in humans.

Fig. 41: *Momordica charantia* var. muricata — fruits are used as vegetable.

Fig. 42: *Mucuna pruriens* — a species with acclaimed aphrodisiac properties; also used in veterinary medicine.

Fig. 43: *Oroxylum indicum* — fruits resemble the hood of a snake and used for snake bite.

Fig. 44: *Peucedanum dhana* var. dalzelii — roots are used in child health care.

Fig. 45: *Plumeria rubra* — flowers are considered as a symbol of purity by the Kondh; often planted in cremation grounds by the Poraja.

Fig. 46: *Pogostemon stellatus* — inflorescence can drive off evil spirits and mosquitoes too.

Fig. 47: *Polyalthia cerasoides* — stump is worshiped as family deity by the Gadaba.

Fig. 48: *Polygonum chinense* — used to cure hand pain.

Fig. 49: *Pterocarpus marsupium* — yields a good timber that is considered apt for making cots.

Figs. 50 & 51: *Pterospermum acerifolium* and *Pterospermum xylocarpum* are used for easing labour pain.

Fig. 52: *Rauvolfia serpentina* — roots are eaten for stomach ache.

Figs. 53 & 54: *Remusatia vivipara* — the "poisonous" tubers are ground and applied on septic wounds.

Fig. 55: *Semecarpus anacardium* — pseudocarps make a delicious food when stalked, dried and fried.

Fig. 56: *Sida acuta* — plants are bundled and used as broom.

Fig. 57: *Solanum torvum* — roots are tied as talisman on the arms of women and children of irritating nature.

Fig. 58: *Thunbergia fragrans* — flowers are believed to increase the yield of toddy when tied to the inflorescence of Caryota urens.

Fig. 59: *Vicoa auriculata* — root paste is used to cure headache.

Fig. 60: *Woodfordia fruticosa* — leaf extract is used specifically to cure wounds near ear.

Fig. 61: *Xylia xylocarpa* — seeds are roasted and consumed.

Fig. 62: Bonda sacred groves stand as islands of vegetation amidst the podu fields.

Fig. 63: Cremation grounds are biodiversity rich spots and harbor both native and exotic species.

Fig. 64: Poraja woman drying steamed Pongamia seed flour for extracting oil.

Fig. 65: Bonda are skilled enough to distill liquor out of any food material. Liquor forms an important source of calories during summer.

Fig. 66: The fruits of *Rubus ellipticus* are compared to the eggs of fishes by the Gadaba and are edible.

Fig. 67: Flowers of *Michelia champaca* are used for adornment purposes by women of all communities. Seed oil is used as a mosquito repellant.

Fig. 68: The thonthor is an indigenous trap used for fishing.

Fig. 69: Stems of *Ichnocarpus frutescens* is one of the cordage sources.

Fig. 70: The grubs of the beetle *Rhyncophorus ferrugineus* are a highly valuable source of protein.

Fig. 71: Ambada — a sweet pickle made from mango pulp.

Fig. 72: *Sida acuta* is an important brush fibre used as broom to clean threshing grounds.

Fig. 73: Kareng is a traditional dress woven by the Gadaba women using fibres of *Fleurya interrupta*. The material shown in the photograph is a remanant piece from a tradition that is now extinct among the San Gadaba.

Fig. 74: Leaves of *Bauhinia vahlii* are commonly used to make meal plates.

Fig. 75: Grains are normally stored in bamboo structures plastered with mud and cow dung.

Fig. 76: Meal plates from leaves of *Bauhinia vahlii* and brooms from Sida acuta are made in excess during seasons and stored on roof tops for later use.

Fig. 77: Streams are usually terraced for rice cultivation.

Fig. 78: Cultural keystone species such as *Mangifera indica* are always left uncut while the area is cleared for podu by the Bonda.

Fig. 79: Olsi is generally cultivated in lower mountain slopes and plateaus.

Fig. 80: Heeya pidika — a traditional cultivar of scented rice conserved by the Kondh community.

Fig. 81: Chipti — a cultivar sown by all communities that breaks less while milling.

Fig. 82: Roktha kosa — a red rice cultivar.

Fig. 83: Batta dhan — a cultivar yielding coarse grains; cultivated mostly in hill slopes.

Fig. 84: Nini / nuni dhan — a cultivar of rice known for its smaller size and less-breakage.

Figs. 85 & 86: Bod sua and mami sua. Sua/suan crop is mostly grown in hill slopes. It requires less manuring and requires little maintenance.

Fig. 87: Dosra budeli — a cultivar of mandia (*Eleusine coracana*). Mandia pejo (porridge) forms an important food of the tribal communities. It is consumed in a watery form as a thirst quencher.

Fig. 88: Agricultural activities involve all the members of a family.
A Kondh family threshing paddy at Kondh Pungar.

Fig. 89: The festival of mag porv happens
in the sacred grove. It initiates the first crop season.

Fig. 90: Bejeyini, the women ritualists of Kondh community, play an important role in the Kondh religion.

Fig. 91: Once in 12 years a cow is sacrificed at the sacred grove during the festival of mag porv.

Fig. 92: Bali porv festival initiates the second crop season for the Kondh community.

Fig. 93: The Bejeyni performing a ritualistic dance as a part of mag porv.

Fig. 94: Flowers of Plumeria *rubra* are used for adornment purposes by the Bejeyni during rituals of mag porv.

Fig. 95: Painting on walls is a part of the festival nondi porv. The painting depict the natural resources of the region.

Fig. 96: Painting on walls is a part of the festival nondi porv. The painting depict the natural resources of the region.

Fig. 97: Beating drums and applying sacred marks on the forehead is a part of the festival nondi porv.

Fig. 98: A black hen is being prepared for sacrifice for Nondi porv.

Fig. 99: Annual Ritual Hunt always begins with rituals performed by the Dishari.

Fig. 100: Shooting an arrow at an egg was a traditional test for testing the archery skills of the men folk that has now turned into a ritual.

Fig. 101: Dances are an integral part of tribal festivals that enhance community bonding.

Fig. 102 & 103: Rituals are performed in agricultural fields to ward off pests.
Such ritualistic structures left behind in the
field harbor colony spiders that trap insect pests besides acting as bird perches.

Fig. 104: Flour balls containing the vital herbs required for brewing.

Figs. 105-111: The series of photographs illustrate the
indigenous process of oil extraction from seeds of Pongamia pinnata.
Extraction of oil from a local resource is a family as well as a community activity.

Fig. 106

Fig. 107

Fig. 108

Fig. 109

Fig. 110

Fig. 111

Bibliography

Abegaz, K., F. Beyene, T. Langsrud & J. A. Narvhus 2002. "Indigenous Processing Methods and Raw Materials of Borde, an Ethiopian Traditional Fermented Beverage." *J. Food Technol. Africa* 7 (2): 59-64.

Adebesin, A.A., N.A. Amusa, & S.O. Fagade 2001. "Microbiological Quality of Locally Fermented Milk (nono) and Fermented Milk-Cereal Mixture (fura da nono) Drink in Bauchi, a Nigerian City." *J. Food Technol. Africa* 6 (3): 87-89.

Adler, D.S., G. Bar-Oz, A. Belfer-Cohen, & O. Bar-Yosef 2006. "Ahead of the Game: Middle and Upper Palaeolithic Hunting Behaviors in the Southern Caucasus." *Curr. Anthropol.* 47 (1): 89-117.

Agrawal, R. 1995. "Sea Change in the Forest: Tribals in Orissa's Eastern Ghats Range are Giving Up Shifting Cultivation and Fiercely Protecting Their Forests." *Down To Earth* July 15: 22-23.

Aiyappan, A. 1948. *Report on the Socio-Economic Conditions of the Aboriginal Tribes of the Province of Madras*. The Commissioner of Museums, Government Museum, Chennai.

Akhter, S., Mohammed Alamgir, Md. Shaw kat I sla Sohel, Md. Parvez Rana, S. J. Monjuru Ahmed & Mohammad Shaheed Hossain Chowdhury. 2010. "The Role of Women in Traditional Farming Systems as Practiced in Homegardens: A Case Study in Sylhet Sadar Upazila, Bangladesh." *Tropical Conservation Science* 3(1) :17-30.

Anonby, E.J. 2006. "Bahendayal: Bird Classification in Southern Luri (Iran)." *J. Ethnobiol.* 26 (1): 1-35.

Austin, D.F. 2007. "Sacred Connections with Cat-tail (*Typha*, Typhaceae) — Dragons, Water-serpents and Reed-maces." *Ethnobot. Res. Appl.* 5:273-303. www.ethnobotanyjournal.org/vol5/i1547-3465-05-273.pdf

Backstrand, J.R., L.H. Allen, E. Martinez & G.H. Pelto (2001). "Maternal Consumption of *Pulque*, a Traditional Central Mexican Alcoholic Beverage:

Relationships to Infant Growth and Development." *Public Health Nutr.* 4 (4): 883-91.

Bahiru, B., T. Mehari & M. Ashenafi 2001. "Chemical and Nutritional Properties of "*tej*", an Indigenous Ethiopian Honey wine: Variations within and between Production Units." *J. Food Technol. Africa* 6 (3): 104-08.

Baliarsing, R. & P.K.Nayak 1996. "Totemic Groups, Title Groups and other Social Groups among the Gadaba." *Adivasi* XXXVI (1&2): 14-25.

Banerjee, S. 1969. *Ethnographic Study of the Kuvi Kandha.* Anthropological Survey of India, Calcutta.

Barnard, A. 2006. "Kalahari Revisionism, Vienna and the 'Indigenous Peoples' Debate." *Soc. Anthropol.* 14 (1): 1-16.

Barrera 1972, cited in Toledo, V.M. 1995. "New Paradigms for a New Ethnobotany: Reflections in the Case of Mexico." In: Schultes, R.E. & S. von Reis (eds), *Ethnobotany: Evolution of a Discipline.* Chapman and Hall, London. pp. 75-88.

Behura, N.K. & K.K. Mohanty 1998. "Vulnerable Ethno-Cultural Groups (VEGs)." *Adivasi* XXXVIII (1& 2): 1-24.

Belcher, M., M. Ruiz-Perez & R. Achdiawan 2003. "Global Patterns and Trends in NTFP Development." Paper Presented at *The International Conference on Rural Livelihoods, Forests and Biodiversity* 19-23 May. Bonn, Germany.

Bell, R.C.S. 1945. *Orissa District Gazetteer (Koraput).* Orissa Government Press, Cuttack.

Berger, P. 2001. "Feeding the Dead: Rituals of Transformation among the Gadaba of Koraput." *Adivasi* 40 & 41: 35-50.

Berlin, B., D. E. Breedlove & P. H. Raven 1974. *Principles of Tzeltal Plant Classification: An Introduction to the Botanical Ethnography of a Mayan-speaking People of Highland Chiapas.* Academic Press, New York.

Berlin, B. 1992. *Ethnobiological Classification — Principles of Categorization of Plants and Animals in Traditional Societies.* Princeton University Press, New Jersey.

Bernard, H.R. 1992. *Preserving Language Diversity: Human Organization* 51(1): 82-89.

Boraiah, K.T., R. Vasudeva, S.A. Bhagwat & C.G. Kushalappa 2003. "Do Informally Managed Sacred Groves have Higher Richness and Regeneration of Medicinal Plants than State-managed Reserve Forests?" *Curr. Sci.* 84 (6): 804-08.

Botanical Survey of India 1988. *Final Report of the All India Co-ordinated Research Project on Ethnobiology, Phase I. Ethnobiology of Andhra Pradesh*. Botanical Survey of India, Coimbatore.

Brian, B.B., M. Ruiz-Perez & R. Achdiawan 2003. "Global Patterns and Trends in NTFP Development." Paper presented at *The International Conference on Rural Livelihoods, Forests and Biodiversity*, 19-23 May 2003, Bonn, Germany.

Brown, K. 1992. "Medicinal Plants, Indigenous Medicine and Conservation of Biodiversity in Ghana." *CSERGE Working Paper GEC*: 92-36. http://www.uea.ac.uk/env/cserge/pub/wp/gec/gec_1992_36.pdf

Brummit, R.K. & C.E. Powell 1992. *Authors of Plant Names*. Royal Botanic Gardens, Kew.

Brush, S.B. 2005. "Farmers' Rights and Protection of Traditional Agricultural Knowledge." *CAPRI Working Paper* no. 36. International Food Policy Research Institute. Washington.

Cavendar, A.P. & M. Alban 2009. "The Use of Magical Plants by Curanderos in the Ecuador Highlands." J. Ethnobiol. Ethnomed. *5:3. http://www.ethnobiomed.com/content/5/1/3.*

Census of India 2001, *Final Population Totals*, 2003. Electronic Version released by the Office of Registrar General, New Delhi, India.

Champion, H.G. & S.K. Seth 1962. *A Revised Survey of the Forest Types of India*. Govt. of India Press, Nasik.

Chaudhury, S.K 1989. "Social Organisation of the Kondhs: Some Preliminary Observations." *Indian Anthropol*. 19 (172): 31-51.

Choudhury, P.R. 2007. "Forest-route to PovertyAlleviation — Myths and Realities: Analysis of NTFP-Livelihood Linkages in Some Indian States." Paper accepted for publication in the *Conference Proceedings of The International Conference on Poverty Reduction and Forests: Tenure, Market and Policy Reforms*, Organised by RECOFTC, 3-7 September 2007, Bangkok.

Cotton, C.M. 1997. *Ethnobotany: Principles and Applications*. John Wiley & Sons, West Sussex, England.

Dafni, A. & E.Lev 2002. "The Doctrine of Signatures in Present-Day Israel." *Econ. Bot*. 56 (4): 328-34.

Dafni, A., E. Lev, S. Beckmann & C. Eichberger 2006. "Ritual Plants of Muslim

Graveyards in Northern Israel." *J. Ethnobiol. Ethnomed.* 2:38. http://www.ethnobiomed.com/content/2/1/38

Dalton, E.T. 1872. *C.S.I., Descriptive Ethnology of Bengal*. Calcutta: Office of the Superintendent of Govt. Printing; Cosmo Publications, New Delhi.

Das, P. 2010. "People's Victory: Vedanta will not be able to Mine the Niyamgiri Hills for Bauxite, and for the Dongria Kondhs the MoEF Decision is a Major Victory." *Frontline*. 27:19. http://www.flonnet.com/fl2719/stories/20100924271902600.htm

Dash, R.N. & M.K. Pradhan 2006 (eds.), *The Art and Crafts of Kondhs*. Council of Professional Social Workers (CPSW), Bhubaneswar.

David, N. & C. Kramer 2001. *Ethnoarchaeology in Action*. Cambridge University Press, United Kingdom. pp. 116-37.

Davidson-Hunt, I. 2000. "Ecological Ethnobotany: Stumbling toward New Practices and Paradigms." *MASA Journal* 16 (1): 1-13.

Davis, W.E. 1995. "Ethnobotany: An Old Practice, a New Discipline." In: Schultes, R.E. & S. von Reis (eds.), *Ethnobotany: Evolution of a Discipline*. Chapman and Hall, London. pp. 40-51.

De Feo, V. 2003. "Ethnomedical Field Study in Northern Peruvian Andes with Particular Reference to Divination Practices." *J. Ethnopharmacol.* 85: 243-56.

Devasahayam, N. 1970-71. "Meriah and Ram-Tribal Rituals in Agriculture." *Adibasi* XII (1-4): 97-99.

Devi, B. 1972. "How and What the Kuvi Kondhs Adorn?" *Adibasi* XIV (1): 21-26.

Devi, S. 2002. *Socio-Economic Condition of a Primitive Tribe: A Case Study of the Didayi of Malkangiri District*. M.Phil. Thesis (Unpublished). Berhampur University. Berhampur.

Down To Earth 1995. "Shortages of Plenty: Local Resource Depletion started when Hunter-gatherer Humans grew Roots." February 15. p. 27.

Elferink. J.G.R. 1999. "Teotlaqualli: The Psychoactive Food of the Aztec Gods." *J. Psychoact. Drugs* 31 (4): 435-40.

Ellis, E.C. & N. Ramankutty 2008. "Putting People in the Map: Anthropogenic Biomes of the World." *Front. Ecol. Environ.* 6 (8): 439-47.

Elwin, V. 1988. *Tribal World of Verrier Elwin: An Autobiography*. Oxford University Press, Delhi.

Eyssartier, C., A.H. Ladio & M. Lozada 2008. "Cultural Transmission of Traditional Knowledge in Two Populations of North-western Patagonia." *J. Ethnobiol. Ethnomed.* 4:25. http://www.ethnobiomed.com/content/4/1/25.

Export-Import Bank of India 2003. *Export Potential of Indian Medicinal Plants and Products*. Occasional paper no. 98. Mumbai.

Forest Survey of India 2003, *State of the Forest Report, 2003*. Ministry of Environment and Forests, Dehradun.

Forsberg, F.R. & H. Sachet 1965. *Manual of Tropical Herbaria*. Regnum Veg. Vol. 39. The Netherlands.

Franco, F.M. & D. Narasimhan 2006. Cultural and Ecological Importance of Sacred Groves of Kondhs. *SDNP Newsletter* 1:2-3. www.ecoheritage.cpreec.org/sdnp_publications_newsletter.html

———, 2008. Ecological Elements in the Folk Songs of the Poraja and the Ancient Tamils. *Indian J. Ecocriticism* 1: 49-54.

Franco, F.M., D. Narasimhan & W. Stanley 2008. "Relationship between Four Tribal Communities and their Natural Resources in the Koraput Region." *Ethnobot. Res. Appl.* 6: 481-85. www.ethnobotanyjournal.org/vol6/i1547-3465-06-481.pdf.

Gadgil, M. & V.D. Vartak, 1975. "Sacred Groves of India: A Plea for Continued Conservation." *J. Bombay Nat. Hist. Soc.* 72 (2): 314-21.

Gadgil, M. & P.R.S. Rao, 1998. *Nurturing Biodiversity: An Indian Agenda*. Centre for Environment Education. Ahmedabad.

Gamble, J. S. 1915-1936. *Flora of the Presidency of Madras*. Adlard & Son Ltd. Hart Street, W.O.

Gessler, M.C., D.E. Msuya, M.H.H. Nkunya , A. Schard, M. Heinrich & M. Tanner 1995. "Traditional Healers in Tanzania: Sociocultural Profile and Three Short Portraits." *J. Ethnopharmacol* 48: 145-60.

Ghosh, P. 2002. "Wild Foods." *Down To Earth*. April 30. pp. 26-27.

Government of Orissa 2008. *State Agricultural Policy*. State Agricultural Department, Government of Orissa. orissagov.nic.in

Guenther, M. 2006. "The Concept of Indigeneity." *Soc. Anthropol.* 14 (1): 17-32.

Gupta, S.M. 1996. *Plants in Indian Temple Art*. B.R. Publishing Corporation, Delhi.

Harshberger, J.W. 1896. "The Purpose of Ethnobotany." *Bot. Gaz.* 21: 146-58.

Henkel, T.W. 2005. "Parakari, An Indigenous Fermented Beverage using Amylolytic Rhizopus in Guyana." *Mycologia* 97 (1): 1-11.

Holman, E.W. 2005. "Domain-specific and General Properties of Folk Classifications." *J. Ethnobiol.* 25 (1): 71-91.

Hunn, E. 1982. "The Utilitarian Factor in Folk Biological Classification." *Am. Anthropol.* New series 84 (4): 830-47.

Indigenous and Tribal Peoples Convention, C169 1989. International Labour Organisation. http://www.ilo.org/ilolex/cgi-lex/convde.pl?C169

Jain, S.K. 1986. "Ethnobotany." *Interdiscipl. Sci. Rev.* 11 (3): 285-92.

Jain, S.K. & V. Mudgal 1999. *A Handbook of Ethnobotany*, Bishen Singh Mahendra Pal Singh, Dehradun, India.

Jain, S.K. 2001. "Ethnobotany in Modern India." *Phytomorphology Golden Jubilee Issue*, pp. 39-54.

Jena, M.K., P. Pathi, J. Dash, K.M. Patnaik & K. Seeland 2002. *Forest Tribes of India*, Vol. 1: The Dongaria Kondh. In: K. Seeland & F. Schmithusen (eds.), Man and Forest Series 2. D.K. Printworld, New Delhi.

Jena, M.K., P. Pathi, K.M. Patnaik & K. Seeland 2006. *Forest Tribes of India*, Vol. 2: The Kuttia Kondh. In: K. Seeland & F. Schmithusen (eds.), Man and Forest Series 6. D.K. Printworld, New Delhi.

Kakudidi, E.K. 2004. "Folk Plant Classification by Communities around Kibale National Park, Western Uganda." *Afr. J. Ecol.* 42 (Suppl. 1): 57-63.

Kala, C.P. & B.S. Sajwan 2007. "Revitalizing Indian Systems of Herbal Medicine by the National Medicinal Plants Board through Institutional Networking and Capacity Building." *Curr. Sci.* 93 (6): 797-806.

Kazembe, T. 2007. "Traditional Medicine in Zimbabwe." *Rose+Croix Journal* 4: 55-72.

King, I.O. 2005. *Sacred forests of Kolli hills, Tamil Nadu, India: A study on Botany, Ecology and Community Interactions*. Ph.D. Thesis (unpublished). University of Madras, Chennai.

Kumar, P. & N. Tarui, 2003. "Does Indigenous Knowledge Contribute towards the Benefits of Bioprospecting?" www.ieg.nic.in/workpk234.pdf

Kunhikannan, C. 2005. "Traditions, Rituals and Biodiversity in Sacred Grove of Karakkakavu, Kasargod district, Kerala State: A Case Study." In: C. Kunhikannan & B. G. Singh (eds.), *Strategy for Conservation of Sacred Groves.*

Institute of Forest Genetics and Tree Breeding, Coimbatore, India, pp. 125-36.

Lahiri, D. 1970. "Dynamic Geology in Prehistoric India and Vedic Literature." *IJHS*. 5 (1): 188-96.

Lefèvre, Gabriel. 2008. "An Anthropological Approach to Therapeutic Strategies for Ethnopharmacology: The Case of South-Western Madagascar." *Ethnobot. Res. Appl.* 6: 29-34. www.ethnobotanyjournal.org/vol6/i1547-3465-06-029.pdf

Lev, E. 2002. "The Doctrine of Signatures in the Medieval and Ottoman Levant," *Vesalius* VIII (1): 13-22. http://www.bium.univparis5.fr/ishm/vesalius/VESx2002x08x01x013x022.pdf

Lingard M., N. Raharison, E. Rabakonandriana, J. Rakotoariosoa & T. Elmqvist 2003. "The Role of Local Taboos in Conservation and Management of Species: The Radiated Tortoise in Southern Madagascar." *Conserv. Society* 1 (2): 223-46.

Mahapatra, K. 1972. "The Concept of Primitive." *Adibasi* XIV (1): 7-10.

Mahapatra, R. 2004. "A Great Sweep: Feisty Women Hill Broom Gatherers Force the Forest Department to Relent in an Orissa Village." *Down To Earth* June 15. p. 44.

Mahapatra, A.K., H.J. Albers & E.J.Z. Robinson 2005. "The Impact of NTFP Sales on Rural Households' Cash Income in India's Dry Deciduous Forest." *Environ. Manag.* 35 (3): 258-65.

Maheshwari, J.K. 1997. "Maintenance and Conservation of 'Heirloom' Varieties in Indian Agro-Ecosystems." In: L. Sperling & M. Loevinsohn (eds.), *Using Diversity: Enhancing and Maintaining Genetic Resources on-farm*. International Development Research Centre, Canada. http://www.idrc.ca/en/ev-85308-201-1-DO_TOPIC.html

Maikhuri, R.K., K.S. Rao & K.G. Saxena. 2003. "Integrated Natural Resource Management: The Concept and its Application for Sustainable Development in the Himalaya." In: P.S. Ramakrishnan, K.G. Saxena, S. Patnaik & S. Singh (eds.), *Methodological Issues in Mountain Research, A Socio-ecological Systems Approach*. Oxford and IBH Publishing Company, New Delhi. pp. 177-98.

Malhotra, K.C., N.S. Hemam, S. Stanley & A. K. Gupta 1998. *Community Initiated Forest Protection and Management in Koraput District of Orissa, India*. http://srdis.ciesin.columbia.edu/cases/india-010.html

Malhotra, K.C., Y. Gokhale, S. Chatterjee & S. Srivastava 2001. *Cultural and Ecological Dimensions of Sacred Groves in India*. Indian National Science Academy, New Delhi & Indira Gandhi Rashtriya Manav Sangrahalaya, Bhopal. http://www.ces.iisc.ernet.in/biodiversity/newsletter/issue4/Yogesh_CEdimensions.pdf

Mandal, H., S. Mukherjee & A. Datta 2002. *India: An Illustrated Atlas of the Tribal World*. Anthropological Survey of India, Calcutta.

Manjula, C. 2004. *Plant Resources: Traditional Knowledge of the Irulas of Northern Tamil Nadu*. Ph.D. Thesis (unpublished). University of Madras, Chennai.

Marglin, F.A. & P.C. Mishra 1993. "Sacred Groves: Regenerating the Body, the Land, the Community." In: W. Sachs (ed.), *Global Ecology: A New Arena of Political Conflict*. Fernwood Books Ltd. Canada. pp. 199-207.

Martin, G.J. 1995. *Ethnobotany: A People and Plants Conservation Manual*. Chapman & Hall. London.

McGhee, K. 2003. *Bioprospecting*. www.amonline.net.au/factsheets/bioprospecting.htm

Mgbeoji, I. 2007. "African Indigenous Knowledge Systems and Patents: Is the Patent System Relevant to the Native Healers of Southern Nigeria?" In: E.K. Boon & L. Hens (eds.), *Indigenous Knowledge Systems and Sustainable Development: Relevance for Africa, Tribes and Tribals* 1: 77-92.

Misra, K.M. 2002. "Kandha Tribes of Orissa : An Ecological Perspective." *Adivasi* 42: 110-25.

Mohanty, B.B. 1976. "The Economic Backwardness of the Gadabas." *Adibasi* XVI (2): 55-57.

Mohanty, S.C. 1992. *Paroja: Tribes of Orissa*. Scheduled Castes and Scheduled Tribes Research and Training Institute, Bhubaneswar. pp. 249-58.

Mohanty, K.K. & N.K. Behura 2005. "Tribes of Orissa: An Inquiry into Anthropological Ecology." *Adivasi* 45 (2): 12-29.

Mohanty, B. 2009. "Bihani Maa, The 'Seed Mothers'." *LEISA INDIA*. 11 (3): 39-40.

Napoli, M. 2008. "The Plants, Rituals and Spells that 'Cured' Helminthiasis in Sicily." *J. Ethnobiol. Ethnomed.* 4:21. http://www.ethnobiomed.com/content/4/1/21

Naranjo, P. 1995. "The Urgent Need for the Study of Medicinal Plants." In: R.E.

Schultes & S. von Reis (eds.), *Ethnobotany: Evolution of a Discipline*. Chapman and Hall, London. pp. 362-68.

Natesh, S. 2001. "The Changing Scenario of Herbal Drugs: Role of Botanists. Trends in Plant Sciences." *Phytomorphology Golden Jubilee Issue*: 75-95.

Nautiyal, S., V. Bisht, K.S. Rao & R.K. Maikhuri 2008. "The Role of Cultural Values in Agrobiodiversity Conservation: A Case Study from Uttarakhand, Himalaya." *J. Hum. Ecol.* 23(1): 1-6.

Nayak, R., B.M. Boel & N. Soreng 1990. *The Kondhs: A Handbook for Development*. Indian Social Institute. New Delhi.

Nayak, P.K. 1992. "Social, Cultural and Linguistic Variations in Tribal Orissa." *Tribes of Orissa*. Scheduled Castes and Scheduled Tribes Research and Training Institute, Bhubaneswar. pp. 1-17.

Newmaster, S., R. Subramanyam, N.C. Balasubramaniyam & R.F. Ivanoff 2007. "The Multi-mechanistic Taxonomy of the Irulas in Tamil Nadu, South India." *J. Ethnobiol.* 27 (2): 233-55.

Ogundele, S.O. 2007. "Aspects of Indigenous Medicine in South Western Nigeria." *Ethno-Med.* 1(2): 127-33.

Okoli, R.I., O. Aigbe, J.O. Ohaju-Obodo & J.K. Mensah 2007. "Medicinal Herbs Used for Managing Some Common Ailments among Esan People of Edo State, Nigeria." *Pak. J. Nut.* 6 (5): 490-96.

Oladeinde, F.O., E.I. Nwankwo, O.A. Moronkola, M.A. Amosu, & B. Farayola 2002. "Determination of Indigenous and Foreign Alcoholic Beverages Levels in Urine by Quantitative Infrared Spectroscopy." *African J. Biomed. Res.* 5: 73-76.

Oxford English Dictionary 2002. Edited by Judy Pearsall, Oxford University Press, New Delhi.

Patnaik, P.S.D. 1972. "Bejunis, Their Initiation into Shamanhood." *Adibasi* XIV (1): 11-20.

———, 1976. "Liquor among Dongria Kondhs of Orissa." *Adibasi* XVI (3): 52-57.

Patnaik, N., B. Chowdhury & P.S.D. Patnaik 1984. *The Bondos and their Response to Development*. Tribal and Harijan Research-cum-Training Institute, Bhubaneswar, Orissa.

Patnaik, T., B.B. Mohanty & P.K. Nayak 2002. "The Dongria Kondh Youth Dormitory: An Agent of Development." *Adivasi* 42 & 43: 17-38A.

Patra, B.C. & S.R. Dhua 2003. "Agro-morphological Diversity Scenario in Upland Rice Germplasm of Jeypore Tract." *Genet. Resour. Crop Ev.* 50 (8): 825-28.

Pearce, J.M.S. 2008. "The Doctrine of Signatures." *Eur Neurol.* 60:51-52.

Pearn, J. 2005. "The World's Longest Surviving Paediatric Practices: Some Themes of Aboriginal Medical Ethnobotany in Australia." *J. Paediatr. Child Health.* 41: 284-90.

Platt, B.S. 1955. "Some Traditional Alcoholic Beverages and their Importance in Indigenous African Communities." *Proc. Nutr. Soc.* 14 (2): 115-24.

Pradhan, K.B. & K.H. Badola 2008. "Ethnomedicinal Plant Use by Lepcha Tribe of Dzongu Valley, Bordering Khangchendzonga Biosphere Reserve, in North Sikkim, India." *J. Ethnobiol. Ethnomed.* 4:22. http://www.ethnobiomed.com/content/4/1/22

Pramukh, K.E.R. & P.E.S. Palkumar 2006. "Indigenous Knowledge: Implications in Tribal Health and Disease." *Stud. Tribes Tribals* 4 (1): 1-6.

Prance, G.T. 1997. "The Ethnobotany of the Amazon Indians as a Tool for the Conservation of Biological Diversity." *Monograf. Jard. Bot. Cordoba* 5: 135-43.

Pushpangadan, P. 1984. *Ethnobiology in India: A Status Report.* All India Co-ordinated Research Project on Ethnobiology. Ministry of Environment and Forests. New Delhi.

Raghavendra, S., C.G. Kushalappa & N.A. Prakash 2005. "An Assessment of Rare, Endangered and Threatened (RET) Medicinal Tree Diversity in Sacred Forests of Kodagu, Central Western Ghats and Their Conservation." In: C. Kunhikannan & B. Gurudev Singh (eds.), *Strategy for Conservation of Sacred Groves.* Institute of Forest Genetics and Tree Breeding, Coimbatore, India. pp. 53-58.

Ramakrishnan, P.S. 2005. "In Harmony with Nature." In: P.S. Ramakrishnan, R. Boojh, K.G. Saxena, U.M. Chandrasekhara, D. Depommier, S. Patnaik, O.P. Toky, A.K. Gangwar & R. Gangwar. *One Sun, Two Worlds: An Ecological Journey.* Oxford and IBH Publishing Company, New Delhi.

Rao, N.R. & A.N. Henry 1996. *The Ethnobotany of Eastern Ghats in Andhra Pradesh, India.* Botanical Survey of India, Calcutta.

Rao, G.V.S. & G.R. Kumari 2002. *Flora of Vishakhapatnam District, Andhra Pradesh.* Botanical Survey of India, Kolkata, Vol. I.

———, 2008. *Flora of Vishakhapatnam District, Andhra Pradesh*. Botanical Survey of India, Kolkata, Vol. II.

Rivera, D., C. Obon, C. Inocencio, M. Heinrich, A. Verde, J. Fajardo & R. Lorach 2005. "The Ethnobotanical Study of Local Mediterranean Food Plants as Medicinal Resources in Southern Spain." *J. Physiol. Pharmacol.* 56 (1): 97-114.

Rousseleau, R. & K. Behera 2002-03. "Scheduled Tribes and Forgotten Kings: Ethnohistory of the Joria Paraja in the Erstwhile Nandapur-Jeypore Kingdom." *Adivasi* 42 & 23: 49-63.

Roy, B., C. P. Kala, N.A. Farooquee & B.S. Majila 2004. "Indigenous Fermented Food and Beverages: A Potential for Economic Development of the High Altitude Societies in Uttaranchal." *J. Hum. Ecol.*15 (1): 45-49.

Ruck, C.A.P. & B.D. Staples 2001. "Mistletoe, Centaurs and Datura." In: C.A.P. Ruck , B.D. Staples & C. Heinrich (eds.), *The Apples of Apollo, Pagan and Christian Mysteries of the Eucharist*. Carolina Academic Press, Durham, North Carolina, pp. 15-30.

Sabat, P., N.C. Das & J. Dash 1999. "Socio-Economic and Demographic Profile of the Bodo Gadaba of Koraput District." *Adibasi* XXXVIII (1&2): 32-38.

Sadgopal, M. 2008. "The Birth Givers: Traditional Midwives in India-Dais and Their Outlook." *Heritage Amruth* 4 (6): 17-19.

Sahoo, A.C. 1996. "The Traditions and Customs of the Dongria Kandha of Orissa." *Adivasi* XXXVI (1&2): 34-39.

Sajem, A.L. & K. Gosai 2006. "Traditional Use of Medicinal Plants by the Jaintia Tribes in North Cachar Hills district of Assam, North-east India." *J. Ethnobiol. Ethnomed.* 2: 33. http://www.pubmedcentral.nih.gov/articlerender.fcgi?artid=1563446

Sarap, K. 2004. *Participatory Forest Management in Orissa: A Review of Policies and Implementation*. Understanding Livelihood Impacts of Participatory Forest Management Implementation in India and Nepal. Working Paper No. 2. Overseas Development Group, University of East Anglia, Norwich.

Sashikumar, C. 2005. "Avifauna of the Sacred Groves of North Kerala." In: C. Kunhikannan & B. Gurudev Singh (eds.), *Strategy for Conservation of Sacred Groves*. Institute of Forest Genetics and Tree Breeding, Coimbatore, India. pp. 97-107.

Satpathy, G.N. 1992. "Orissa Tribes and their Homeland." *Tribes of Orissa*. Scheduled Castes and Scheduled Tribes Research and Training Institute, Bhubaneswar. pp. 21-42.

Saxena, H.O. & M. Brahmam 1994. *The Flora of Orissa*. Regional Research Laboratory and Orissa Forest Development Corporation Ltd. Bhubaneswar. Vol. I.

Saxena, N.C. 2003. *The Rural Non-farm Economy in India: Some Policy Issues*. NRI Report No. 2752. Natural Resources Institute, The University of Greenwich. London.

———, 2003 (b). *Livelihood Diversification and Non-Timber Forest Products in Orissa: Wider Lessons on the Scope for Policy Change?* Working Paper 223. Overseas Development Institute, London.

Schultes, R.E. 1962. "The Role of Ethnobotanist in the Search for New Medicinal Plants." *Lloydia* 25 (4): 257-66.

Schultes, R.E. & S. von Reis 1995. "Historical Ethnobotany." In: R.E. Schultes & S. von Reis (eds.). *Ethnobotany: Evolution of a Discipline*. Chapman and Hall, London.

Senapati, N. & N.K. Sahu 1966. *Orissa District Gazetteers-Koraput*. Orissa Government Press, Cuttack.

Senapati, N. & D.C. Kuanr 1984. *Koraput District Gazetteer-Supplement*. Gazetteers Unit, Revenue Department, Government of Orissa.

Sensarma, P. 1998. *Ethnobiological Information in Kautiliya Arthasastra*. Naya Prokash, Calcutta.

Silori, C.S., M. Mamta, M.A. Khalid & V. Paul 2005. "Non-Timber Forest Products: Conservation Status and Management Priorities in the Community Managed Forests of Andhra Pradesh, South India." *Int. J. Sust. Dev. World* 12 (3): 334-46.

Simpson G.G. 1961. *Principles of Animal Taxonomy*. New York: Columbia University Press.

Singh, K.S. 1998. *India's Communities*. Anthropological Survey of India & Oxford University Press. New Delhi.

Sofowara, A. 1982. *Medicinal Plants and Traditional Medicine in Africa*. John Wiley and Sons. Chichester.

Souza, S. P. & A. Begossi 2007. "Whales, Dolphins or Fishes? The Ethnotaxonomy of Cetaceans in Sao Sebastiao, Brazil." *J. Ethnobiol. Ethnomed.* 3 (9). http://www.ethnobiomed.com/content/3/1/9

Stephens, C., J. Porter, C. Nettleton & R. Willis 2006. "Indigenous Health 4: Disappearing, Displaced, and Undervalued: A Call to Action for Indigenous Health Worldwide." *Lancet* 367: 2019-28.

Swain, J. 1968. "The Kondh Pantheon and Their Social Significance." *Adibasi* X (3 & 4): 33-39.

TERI 2004. *Sustainable Exploitation, Cultivation and Marketing Linkages of Non-Timber Forest Products (NTFPs) in Andhra Pradesh*. Report Prepared for the Andhra Pradesh Forest Department. TERI Project Report No. 2003 SF 63.

The World Bank 2006. *India: Unlocking Opportunities for Forest Dependent People*. Oxford University Press. New Delhi.

Thurston, E. 1975. *Castes and Tribes of South India*. Vol. II (Repr. Ed). Cosmo publication, Delhi. pp. 242-52.

Thusu, K.N. & M. Jha 1972. *Ollar Gadaba of Koraput*. Anthropological Survey of India, Calcutta.

The Times of India, August 2009. "Orissa now Odisha, Oriya becomes Odia." http://timesofindia.indiatimes.com/india/Orissa-now-Odisha-Oriya-becomes-Odia/articleshow/5154302.cms. Accessed on 10-09-10.

Trager, J. 1982. "Cradles of Civilization and the Hands that Rocked Them." In: Kirk, D. & K. Eliason (eds.), *Food and People*. Boyd & Fraser Publishing Company, San Francisco. pp. 32-33.

Tsehaye, Y. & F. Kebebew 2006. "Diversity and Cultural Use of Enset (*Enset ventricosum* (Welw.) Cheesman) in Bonga *in situ* Conservation Site, Ethiopia." *Ethnobot. Res. Appl.* 4:147-57. www.ethnobotanyjournal.org/vol4/i1547-3465-04-147.pdf

Ulaka, R.C. 1976. "The Social life of the Paraja." *Adibasi* XVI (2): 21-32.

UNESCO. 2003. *Language Vitality and Endangerment*. Document submitted to the International Expert Meeting on UNESCO Programme Safeguarding of Endangered Languages, Paris, 10-12 March 2003.

UNESCO. 2010. *UNESCO Interactive Atlas of the World's Languages in Danger*. http://www.unesco.org/culture/ich/index.php?pg=00206. Accessed on 23-04-10.

Vandebroek, I., J.B. Calewaert, S. De jonckheere, S.Sanca, L. Semo, P.V. Damme, L.V. Puyvelde & N.D. Kimpe 2004. "Use of Medicinal Plants and Pharmaceuticals by Indigenous Communities in the Bolivian Andes and Amazon." *Bull. World Health Organ.* 82 (4): 243-50.

Verma, R.C. 1990. *Indian Tribes through the Ages.* Publication Division, Ministry of Information and Broadcasting, Govt. of India. pp. 5-12.

Vidyarthi, L.P & B.K. Rai 1976. *The Tribal Culture of India.* Concept Publishing Company, New Delhi.

Viswanathan, M.B., E.H.P. Kumar & N. Ramesh 2006. *Ethnobotany of the Kanis.* Bishen Singh Mahendra Pal Singh. Dehradun.

Voeks, R.A. & S. Nyawa 2006. "Dusun Ethnobotany: Forest Knowledge and Nomenclature in Northern Borneo." *J. Cult. Geogr.* Springl Summer. 23 (2):1-31.

Walters, A.H. 1973. *Ecology, Food and Civilisation.* Charles Knight & Co. Ltd. London.

Wasson, G.R. 1986. "Persephone's Quest." In: G.R. Wasson, S. Kramrisch, J. Ott & C.A.P. Ruck (eds.), *Persephone's Quest, Entheogens and the Origin of Religion.* Yale University Press. New Haven and London, pp. 17-81.

General Index

Index

(of Botanical Names)

Index

(of Local Names)